SPUDDING IN

SPUDDING IN

*Recollections
of Pioneer Days
in the California
Oil Fields*

by William Rintoul

Library of Congress Catalog Card Number: 75–28788
ISBN 0–910312–37–0
California Historical Society, 2090 Jackson Street, San Francisco 94109

Published 1976
Printed in the United States of America

Editorial Director: Henry Mayer
Designer: Harlean Richardson
Photo Researcher: Gerda Ray
Executive Editor, Publications, California Historical Society: Marilyn Ziebarth

Graphics reproduced on pages 37, 123, and 127 reprinted, by permission, from American Petroleum Institute, *History of Petroleum Engineering* (1964) and graphics on pages 129 and 181 from American Petroleum Institute, *Finding and Producing Oil* (1939).

Graphics reproduced on pages 19, 49, 50 with the kind permission of The Bancroft Library, University of California, Berkeley.

Graphics reproduced on pages 9, 58, 82, 88, 89, 105, 114, 130, 134, 136, 143, 151, 208 with the kind permission of The Huntington Library, San Marino, California.

The material recorded in the following pages is drawn from the voluntary contributions of men who were personally engaged in various operational phases of the California petroleum industry for diverse periods during the interval from 1890 to 1940. This fifty-year span was by all measures one of the most exciting in petroleum history. Stimulated by two world wars and a burgeoning interest in research, great strides were made. Haphazard drill site selection gave way to scientific exploration and logical choice; drilling changed from an adventure of strong-armed giants to a highly instrumentalized, precision process; storage emerged from the oil sump stage to efficient, conservation-oriented systems; transportation by land and sea increased enormously in both volume and speed, owing to the appreciation and application of the proper engineering principles; and, natural gas, the nuisance product long flared to the atmosphere, became one of the world's most wanted energy products.

This book is the partial story of these eventful years and the men responsible for the changing scene. It is incomplete, of course, but it represents the personal recollections of men who had their day in this important era. Many of them have now passed on, and others are in their twilight years. It is just a pity that many more passed into limbo without leaving for future generations the records of their experiences and adventures.

Richard Sneddon

To Sam Grinsfelder and Dick Sneddon
two dedicated Petroleum Production Pioneers

Contents

❧ *This book has been made possible by
the generous contributions of:*

The R. C. Baker Foundation

Atlantic Richfield Company

J. E. (Brick) Elliott

The Hillman Companies
in memory of Harry H. Hillman

Fritz Huntsinger, Sr.

Earle M. Jorgensen Company

Ross McCollum

Thomas P. Pike Foundation

Republic Supply Company of California

Henry Salvatori

Santa Fe Drilling Company

Standard Oil Company of California

Union Oil Company of California

Francis E. Vaughan

Foreword

For all who know California, the most commonly repeated part of its history is the Gold Rush experience, with its legends and stereotypes emphasizing the acquisitive energies of Californians in extracting treasure from the earth. For the economic growth and character of California today, however, another buried resource—petroleum—is far more important; the cumulative value of the state's oil production is ten times greater than its gold, and California ranks second only to Texas among the nation's oil-producing states. Yet, ironically, the early history of California oil production is not at all well known.

In California's earliest human epoch Native Americans gathered pitch and coal tar from open seeps of asphaltum in the state's canyons and hilly outcroppings. Early European explorers left numerous descriptions of the natural occurrence of oil and its residues. But it was not until the 1860's that California experienced brief oil-drilling booms, first in the vicinity of Santa Paula and later in the Pico Canyon area of Los Angeles County. In these and other flurries of oil excitement all over the state, including the optimistic hamlet of "Petrolia" near Cape Mendocino in Humboldt County, Californians sensed the dramatic fortunes that could be built upon "black gold."

Just as gold mining technology underwent a rapid transformation from the simple pan, rocker, and sluice of 1850's placer-mining to the vast technology required by deep mining in the Comstock Lode in the 1870's, so, too, did oil technology move from the crude simplicity of digging by hand with picks and shovels to the use of power-driven tools, towering derricks, and intricately contrived drilling rigs that probed into geologic mysteries thousands of feet below the surface. This transformation took place in the decades from 1870 to 1940 and is thus within the range of living memory; the oil pioneers, unlike their Gold Rush counterparts, have not entirely disappeared.

To know the Gold Rush experience we turn to the hundreds of published diaries and letters of the miners. Perhaps similar records of the oil fields and their boom towns will yet be discovered. But it is unlikely. The era of diary-keeping had passed by the time the pioneers of oil went to work, and the sense of making history was no longer a personal concern, to be communicated "back home." To know what it was like at Kern River or Signal Hill, to experience vicariously the personal drama and dangers, the great expectations of wildcatters, and the genius of mechanical innovators, we have to turn to the men themselves and listen to their anecdotes and recollections of those remarkable decades.

This book seeks to tell their story, largely in their own words as recounted to interviewers sent out by the Petroleum Production Pioneers of California, an organization of 2,700 members. In 1963 this organization established a folklore committee "to collect and preserve historical facts pertaining to the drilling and production activities of the petroleum industry and the individuals engaged therein." Sam Grinsfelder, a prominent and widely respected Union Oil executive, agreed to serve as chairman of the committee. Through tape recordings and carefully prepared transcripts of conversations with thirty-five pioneers, Grinsfelder and his associates obtained an invaluable record of experience not to be found in the more formal histories published by the larger petroleum companies.

In all, the transcripts of the thirty-five interviews ran to hundreds of pages. This material was turned over to William Rintoul, for twenty years a well-known writer of petroleum history and contemporary oil industry developments. Working with the raw material of these often rambling reminiscences of many events, people, and places, Rintoul (with the editorial assistance of Henry Mayer of San Francisco) has created a narrative of personal experience which, along with the photographs and other illustrations accompanying it, captures the spirit and tempo of life in California's oil industry, 1890–1940.

"Spudding in" is an oil field term for the start of a well-drilling operation. So, too, is this book a beginning. It is by no means intended to be a comprehensive history of the California oil industry. Much that is here constitutes personal memory and opinion, history from the point of view of the participants alone. Yet, these stories are an irreplaceable source for understanding the beginnings of the oil industry in California, for sharing the excitement, and for admiring the ingenuity and fortitude of those times.

J. S. Holliday

EXECUTIVE DIRECTOR
CALIFORNIA HISTORICAL SOCIETY

Acknowledgements

This book would not have been possible without the contributions of Sam Grinsfelder and Richard Sneddon, chairman and vice-chairman, respectively, of the Folklore Committee of the Petroleum Production Pioneers. Arthur Turman and Ronald W. Heath also made important contributions to the committee's work, as did the members of the Desk and Derrick Club of Los Angeles who volunteered to do the stenographic work for the project.

Thirty-five pioneers shared their reminiscences with the Folklore Committee, as follows: Neal H. Anderson, Marion Arnold, R. C. (Carl) Baker, Paul D. Barton, E. L. Bendiger, C. L. Case, Clifford Davis (material provided by Richard Hathaway), Fred E. (Fritz) Davis, Earl M. Delaney, J. E. (Brick) Elliott, Walter English, Oliver C. Field, E. G. Gaylord, Ronald Heath, A. J. L. Hutchinson, Joseph Jensen, Fritz Karge, Warren H. Kraft, Lindsay B. (Lin) Little, Kenneth M. Manley, Ross McCollum, Roy P. McLaughlin, Graydon Oliver, J. W. Pauson, Jack Reed, Kenyon Reynolds, Charles Scharpenberg, R. R. (Dick) Smith, Richard (Dick) Sneddon, Robert Tulin, Arthur Turman, George Tyler, Harold M. Van Clief, Frank Vaughan, Leland K. Whittier, and Garth L. Young. Although not all of the material supplied fit into the scope of the book that has emerged, the contributions of these men form the essence of what is presented in these pages.

I should like to thank the reference staff of Beale Memorial Library, Bakersfield, for its help in making research materials available to me, particularly Nina Caspari, who is in charge of the library's Historical Collection, and Mary Haas, who is in charge of the Geological, Mining, and Petroleum Collection. I should also like to thank Richard C. Bailey, director of the Kern County Museum, Bakersfield, and his staff, as well as Audrey B. Acebedo of the R. C. Baker Memorial Museum, Inc., Coalinga, for their help in furnishing illustrative material. Many other individuals, companies, and organizations also loaned photographs for use in this book; their names appear in the photograph captions. I would also like to offer public thanks to Guy F. Miller (my father-in-law) for his help with information on rotary drilling; he spent his entire working life working on rotary rigs, beginning in 1921 with the Santa Fe Springs boom.

At the California Historical Society I am grateful to Dr. J. S. Holliday, executive director, for his editorial help and his persevering efforts to raise the necessary funds for publication. I also appreciate the important contributions made by Henry Mayer, who worked as editorial director; Marilyn Ziebarth, executive editor for the Society's publications; Harlean Richardson, the book's designer; Gerda Ray, photo research; Peter Bray and Martha Hairston, technical artists. To all I extend my thanks.

William Rintoul

BAKERSFIELD, CALIFORNIA
JULY, 1975

SPUDDING IN

1 San Joaquin Oil and the Kern River Field

Emigrants passing through the San Joaquin Valley in the 1850's occasionally saw burning oil seepages in the foothills that formed the western boundary of the valley. As the existence of oil thus became known to the early settlers, they began to use this "pitch" or "coal tar," as they called it, to lubricate farm machinery. One homesteader, Chris Danielson, who came from Denmark with his young bride to settle on land near the present site of Coalinga, used to gather his family together, load the wagon with empty buckets and a picnic lunch, and head for the nearest oil seep. The family would spend the day collecting buckets of oil. When the family returned home, Danielson's daughter later recalled, her father would use the oil as a substitute for axle grease.

Even before the settlers, the Yokut Indians who inhabited the valley made use of oil. At Tulamniu, near the present site of Taft, Yokut villagers obtained asphalt from nearby seeps and used it in a variety of ways: cementing bundles of yucca fiber to make paintbrushes, coating baskets, cementing chipped scrapers and cutting tools, setting inlays of shell, and securing basket hoppers to the rims of stone mortars. The Yokuts may even have stockpiled oil, according to the findings of the Smithsonian Institution's excavation of the site of Tulamniu in 1933–34. Excavators found small balls of asphalt, some as large as two and one-half inches in diameter, which the Smithsonian's scientists speculated might have been set aside for use as adhesive.

At no other place on the valley's west side were oil sands so largely exposed as in the vicinity of what is now the town of McKittrick, thirty-five miles west of Bakersfield. There beds of asphalt lay visible, oozing crude petroleum. In the 1860's San Joaquin Valley oil activity began with the advent of Mother Lode miners, who sought to quarry the asphalt in great quantities. They used picks, shovels, crowbars, and general mining techniques to open the rock seams in order to produce a free flow of the thick, tarry substance.

Prospectors dug out shallow wells eight to ten feet deep.

Reflections in oil, Kern River, 1901.
(Kern County Museum)

Oil prospectors first dug open pits in areas where asphalt outcropped, as shown at left, near McKittrick. (Kern County Museum)

Prospectors tried to adapt mining methods to petroleum production, as seen in the McKittrick mine, at right, shored with redwood timbers. The author took this photograph in 1956.

The short pieces of lumber they laid down to stand on while shoveling became mired in the tar, and rising gases from the seeps often made it impossible to continue work. "We could only get to a depth of about eight feet, because then the oil and sand began to run into the hole, and we could not stand in it," John L. Sullivan, a pioneer San Joaquin oil prospector, told F. F. Latta in 1939.* "A dozen of these pits were dug, and as fast

* Recollections of the earliest San Joaquin oil pioneers may be found in F. F. Latta, *Black Gold of the Joaquin* (Caldwell, Idaho; 1949). Historians of California will long be indebted to F. F. Latta for the decades of painstaking effort he has put into collecting, preserving, and publishing the stories of California pioneers and their descendants. We gratefully acknowledge the use in this chapter of brief quotations from interviews first published in *Black Gold of the Joaquin*, as follows: John L. Sullivan (pp. 43–44), Keny Pool (pp. 65–66), H. A. Blodget (pp. 79–80), Angus Crites (p. 102, 112), Dora Mc-Whorter Banta (p. 157), Alfred Harrell (p. 174) and Milton McWhorter (p. 175).

as the oil would seep into them, we would bail it out with buckets and put it into barrels on the light wagon. I guess what I am calling oil was really a light asphaltum, because it was thick and gummy and not like the oil we see today coming out of deep oil wells."

In 1866 Buena Vista Petroleum erected a still at a spring three miles northwest of McKittrick. Miners took tar-like oil from pits and open cuts to feed the still, which had a capacity of 300 gallons. Sullivan recalled to F. F. Latta memories of "the smoke and mess made by the still. The fumes were bad, and there was much refuse from the kettle—bones, gravel, sand, and thick tar." Buena Vista produced about 3,000 gallons of refined oil, but the expense of transporting the product to market doomed the enterprise to failure. The same company later erected a three-kettle refinery to treat the superficial deposits of asphalt, but this enterprise also failed.

In 1878 Bill Tibbett and Keny Pool tried to drill a well for Columbian Oil alongside an oil seepage near Reward. They used a hand rig, consisting of a drill string of hickory poles suspended on a tripod with an auger on the end and a wind-lass to raise and lower the bit. But as Pool told F. F. Latta in the 1930's, "We couldn't do a thing with the outfit. The bit would twist into the asphaltum and stick, and we couldn't turn it around. When the bit stuck, the only way we could get it out was to hang a sack of dirt on the handle of the windlass and let it stand for a couple of hours." That method slowly pulled the bit loose, but Pool ruefully observed that they had previously made "a deeper hole digging by hand."

Instead of the portable tripod, Tibbett and Pool decided to build a stationary derrick. Using a much passed-around picture of a Pennsylvania derrick, they built a forty-five-foot high rig, which Pool told F. F. Latta was the first permanent derrick in the San Joaquin Valley. The derrick allowed for more leverage, but Pool and Tibbett pushed the auger bit around themselves by means of a crossbar which passed through the top joint of the poles at chest level. By this method they attained a depth of forty feet before the asphaltum became too sticky and the effort was abandoned.

In 1893 the Southern Pacific Railroad extended a spur line to the community of McKittrick, then known as Asphalto. The railroad, in association with Solomon Jewett and Hugh S. Blodget of Bakersfield, organized a company, Standard Asphalt, and built a twenty-one-kettle refinery near the railhead. Three hundred tons of crude asphalt produced about 100 tons of refined asphalt, which was priced at an average of $25 per ton. Miners sunk two shafts on vein-like deposits to depths of 280 and 310 feet, using a horse to power a hoisting windlass and a gasoline engine to force air into the mines.

Elva Coulthard, whose father was a carpenter and blacksmith for Standard Asphalt, later recalled that the "town" of Asphalto consisted of the refinery, the superintendent's home and office (which also served as the post office), the bunkhouse, the cookhouse, and one saloon. "There were no oil wells around in those days," she said, "except the ones that were dug by manpower to obtain some oil to use in refining the asphalt. Some crude oil was added to the asphalt chunks in the huge vats. Then it was melted and run into boxes somewhat smaller than an apple box. It was shipped in this form

Jewett & Blodget operations in the San Joaquin Valley; (left) Maricopa, 1894; (above) the Tevis well, Sunset, 1895. (Standard Oil Company of California).

7

and used for paving streets. A small amount of crude oil was refined, and the resulting 'distillate', as it was called, was used to run an engine which operated a dynamo that furnished electricity for light," possibly the first electric lights in Kern County.

"When anyone tried to bore a well for water in that vicinity," Elva Coulthard said, "they always struck oil. All the water we used was brought by rail in tank cars, and we paid a dollar a barrel for it. Oil was so abundant it oozed out of the hillside, mixed with dirt and sagebrush, and hardened. This was broken up and used for fuel in the stoves, like coal."

Meanwhile, Jewett and Blodget were showing interest in oil seeps in the Old Sunset area near the later site of Maricopa, twenty miles southeast of Asphalto. Activity had begun in 1889 when Jewett, Blodget, John Hambleton, Judge J. O. Lovejoy, J. H. Woody, William F. Woods, and others located 2,000 acres along the edge of the hills. They organized Sunset Oil and started a well in a bed of *brea* just at the point where oil sands outcropped. They got a strong flow of sulphur water. A second well found heavy oil but was abandoned. When the third and fourth wells proved unsuccessful, other investors pulled out, leaving Jewett and Blodget to pursue the development. Blodget drilled a number of shallow holes, getting small amounts of oil which were enough to supply the flux for making asphalt.

In 1891 the Jewett & Blodget refinery was established at Old Sunset. The natural asphalt was quarried as at McKittrick. In 1934 H. A. Blodget recalled the scene for F. F. Latta. "The miners worked stark naked, covered with the liquid asphaltum," Blodget said. "At the end of tour they were scraped with a case knife, or the wooden scrapers used on race horses, and washed in distillate." This clean-up was not practical at lunchtime, and Blodget described how "the boys sat in the camp mess 'au naturel,' clothed only in asphaltum that shown like lacquer. They covered the benches with newspapers, which stuck to them like sticky fly paper" and did mock war dances, "newspapers waving in the wind," as they returned to work.

The quarried asphalt was melted in open kettles with a

Prospectors drilled first by oil seeps. Ralph Arnold photographed this one at La Brea Rancho, six miles west of Los Angeles, in 1906. An oil derrick shares the skyline with the trees. (Huntington Library).

small amount of crude oil as a flux. Then hot asphalt was drawn off into wooden boxes and shipped to the railhead at Bakersfield by teams of sixteen to twenty-four horses. The teams hauled back barrels of water, staves and hoops, groceries, hay, and feed. Angus Crites, who worked for Jewett & Blodget in the 1890's and later pioneered his own oil companies in the valley, recalled to F. F. Latta that two and a half miles an hour was good time for a freight team, which stayed on the road ten hours a day. Drinking water could only be obtained at two places on the Sunset-Bakersfield route, and "the water at those places tasted as if it were poison," Crites recalled. Drivers of the wagon trains took great pride in making U-turns with their large teams on Chester Avenue in

Bakersfield, drawing sizeable crowds and large wagers. Hank Wren and his team were "the most spectacular sight in Bakersfield during the 1890's," remembered Crites.

In 1898 Milton McWhorter, a veteran oil prospector, built a small refinery at McKittrick, where he manufactured paint, axle grease, and other compounds from crude oil. McWhorter and his associates formed El Dorado Oil and drilled several unsuccessful wells. McWhorter's son, Horace, helped in his father's drilling operations, as did his young daughters, Ella May and Dora, who rode the horse, studying their lessons as the horse went round and round the rig to furnish the drilling power. "I can remember the period of 'spudding in' and setting up the derrick and getting the standard tools to working," Dora McWhorter Banta wrote in a letter to F. F. Latta many years later. "The drillers would, now and then, let me put my hands on the cable as it worked the bit up and down." The McWhorter children made playtoys of asphaltum, and Horace built his young sisters a miniature derrick. The girls often rescued young lambs that had fallen into the asphaltum seeps. Dora also remembered her father's numerous experiments with refining oil in their kitchen and the disastrous fire that once burned down the family home.

The work at McKittrick attracted interest, and Judson F. Elwood, while living in Fresno County, bought a few shares in an oil property there and went down to inspect it. After visiting in McKittrick, he went into Bakersfield to see his brother, James Munroe Elwood, who was running a small wood yard. While talking, Judson glanced in the direction of Kern River and remarked to his brother, "It looks up there something like the country out toward McKittrick, and there might be oil out there too, and you'd better investigate a little."

In a letter published in *California Oil World* in 1911, Jonathan Elwood, father of the two Elwood brothers, recalled what happened. "After this conversation, James Munroe began making some inquiries and soon happened to overhear two drunken fellows talking and one swearing that he knew where there was an oil prospect, up the river on Thomas Means' ranch."

From the earliest days of settlement, people had known of

Hand-dug hole braced with redwood timbers from which Jewett & Blodget produced oil in the 1890's near Maricopa. The author took this photograph in 1974.

an oil seepage at the edge of the Kern River on Means' property, about seven miles north of Bakersfield. Sheepmen in the 1850's were said to have used tarry oil from the seep for branding their stock. James Munroe Elwood went to see Tom Means and found the latter convinced there was oil on his property. Means told Elwood that Kern County Land had graded a ditch years before, under the end of a high bluff close to the river, and that a small seepage of oil had then been seen. The younger Elwood wrote immediately to his father, asking him to come down from Fresno County and help in the search for oil.

The elder Elwood recalled, "He, James Munroe Elwood, and I, Jonathan Elwood, alone, without the assistance of any one, discovered the oil on the north bank of Kern River, seven miles northeast of Bakersfield on Thomas A. Means' farm. This was in May, 1899. We made the discovery with a hand auger, under the edge of a cliff by the river."

The auger consisted of a piece of thin steel about four

inches wide and twisted so as to bore a hole about three inches in diameter. Elwood recalled that they used "a short piece of one-half inch iron rod making the bit and rod together four feet long. A screw was cut on the end of this rod to receive one-half inch gas pipe which we had cut in four and eight-foot lengths, so we could bore one and the other alternately and never have our auger handle more than four feet high above the ground. We bored a number of holes fifteen or twenty feet deep and every time would bore into water sand that we could not keep on our auger."

The Elwoods concluded that "the bank must have slid down and that we were boring where the river had once been. We then went where the bank was worn off by the river perpendicularly thirty feet. We there dug back into the bluff as if making a tunnel for three or four feet and set our auger on solid formation, and in three hours we were in oil sand, at a depth of only thirteen feet. We had enough auger stem with us to go on to the depth of twenty-five feet and it was looking well.

"We then went up onto the bluff and commenced a shaft, and at the depth of forty-three feet we again struck the oil sand. We were then obliged to get timber and curb, as we went down, as the oil sand was too soft to stand up. We were obliged to put in an air blast to furnish fresh air to the man below, on account of the strong odor of gas. At the depth of seventy-five feet, there was so much oil and gas that we concluded we had better get a steam rig."

The Elwoods hired Milton McWhorter as their driller. "The shaft furnished us with oil to run our own steam rig," Jonathan Elwood wrote, and about May, 1899, the first drilled oil well on the Kern River was completed. The well began flowing at about two barrels a day and later increased to fifteen. In quick succession four more wells were drilled, and the boom was on. "The first oil taken away was when I took four whiskey barrels of it to Kern City and shipped it to Millwood for skid grease, getting a dollar a barrel net," Jonathan Elwood recalled.

While the Elwoods were drilling the first well, a noted oil man, E. L. Doheny, of Los Angeles, came to investigate the

A crew poses in front of an oil tank in the Kern River field, and a few fellows find a comfortable spot on the pipe. (Kern County Museum)

operation. He asked the Elwoods if they had any objection to his buying out Means, subject to their lease. When they said they had no objection, Doheny said, "Go over with me, boys, and introduce me and if I make the purchase, I will give you boys $2,500."

Jonathan Elwood recalled in the letter published in *California Oil World* that Doheny "made the trade and about two weeks after this, the boys met with him in Bakersfield. Without a word in regard to his promise, he said, 'Oh, yes, boys, about that $2,500. Here is a check for the amount.' This, of course, was an agreeable surprise to us."

Bakersfield people remained skeptical of the discovery for several months. Although the first Elwood well had come in around the early part of May, 1899, no story appeared in Bakersfield's *Daily Californian* until early June. Alfred Harrell, the owner and editor of the newspaper, explained to F. F. Latta that "none of us thought there was any oil on Kern River . . . .

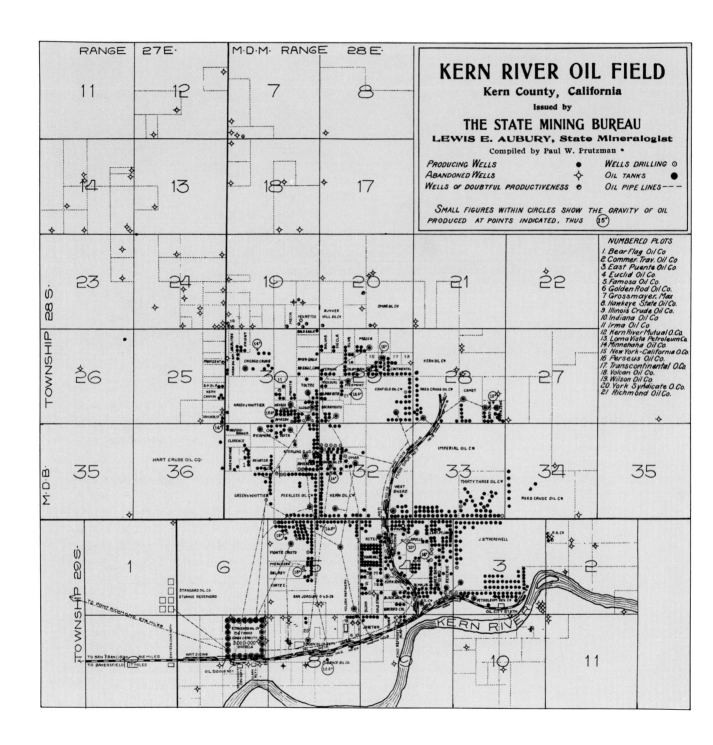

When we heard that someone had gone up there and was digging into the riverbank we considered it as a stock selling scheme and gave it no publicity." Milton McWhorter recounted to Latta that Angus Crites finally persuaded people in Bakersfield that the discovery was genuine. "Crites came to the Elwood well and looked the place over very thoroughly," McWhorter said. "We had whiskey barrels of oil and oil in milk cans, kerosene cans, beer kegs, and everything that could be obtained. He watched us take about four barrels of oil from the dug well with the level remaining the same. Then he went into Bakersfield and told the newspaper men that they had better go up the river and get themselves some oil land, because we surely had an oil field. That was what brought out the first news item."

Hundreds of men came from all over the West to see the Elwood discovery well. At one time a picture was taken of two hundred men in the lobby of Bakersfield's Southern Hotel. An agent for the Southern Pacific bought oil from the Elwoods for use in the line's locomotives. Crude oil soon came out of the Kern River field and left Bakersfield in tank cars by the trainload. For the first time on the West Coast a vast supply of petroleum light enough to be used as fuel had been discovered close to a railroad leading to the great harbor at San Francisco. In the rush, lasting some twenty months, to put down shallow wells, Bakersfield became a boom town.

Small companies proliferated, and almost every issue of *The Daily Californian* reported the formation of another oil company. Their names embodied everything from high hopes to pride in regional origins, and included such companies as the Prosperity and the Blue Bird, the Pennsylvania and the Vancouver, the Clarence, Bald Eagle, Moneta, Hanford-Sanger, Southern Cross, Banner, Berry, and New York-California.

The value of land shot up in the booming oil field. In 1901, when producers shipped almost 12,000 barrels per day from the field, the county assessor announced that he would assess oil lands at $1,000 an acre. The field abounded with stories like that of E. M. Roberts, reported to have leased forty acres on Section 8, T29S-R28E, to Consolidated Crude for a handsome $5,000 cash bonus with one-fifth royalty. The lease

State Mining Bureau map of the Kern River oil field, 1903.

carried a clause giving the oil company an option to purchase the land for $2,500 an acre, or a magnificent total of $100,000. Fifteen years before, Roberts had acquired the parcel as part of eighty acres for which he had traded thirty cows.

Boom times encouraged extravagant talk and ideas. A townsman named Wellington Canfield returned from a six-month trip to the East with the heady news that in each of the "hundreds" of places he visited, people knew and talked about Bakersfield. "Oil has made this city famous," Canfield declared. So famous had the town become that the Southern Pacific promoted a sightseeing excursion from San Francisco.

Turn-of-the-century Bakersfield; a quiet agricultural town before oil operators began to develop the Kern River field and gathered at the Southern Hotel, right foreground. (Kern County Museum)

For a round trip fare of $10.60 tourists could visit Bakersfield and the Kern River oil field.

One San Franciscan, J. W. Pauson, traveled to Bakersfield out of more than idle curiosity. Pauson was the secretary of San Francisco's Central Power and Light Company, a private corporation which supplied gas and electricity to the Emporium, a newly opened department store, and other San Francisco enterprises. As the company's business grew, Pauson recalled, "troubles grew along with it." Central Power and Light burned Welch anthracite, a hard coal, under its boilers. Unexpectedly, the suppliers notified C.P.L. that no more anthracite coal could be obtained. "Our only alternative," said Pauson, "was to burn Wellington Screenings, a soft coal. Soot, cinders, and smoke from our chimney created the first air pollution problem in San Francisco. Our neighbors complained, and I was summoned before the Board of Supervisors and told to abate the nuisance."

The firm had to find other sources of fuel in order to stay in business. "In looking around," recalled Pauson, "I learned that oil was being burned in boilers in the Kern River oil fields. I hurriedly made a trip to Bakersfield—if travel in those days could be termed 'hurried'—and inspected the installation. It was just what we needed, so I arranged with an oil company to send us up three tank cars of fuel oil." The shipment arrived on a cold December day. The heavy oil was like molasses. The tank cars had no heat, and it took some time to unload the oil. At the plant, the light and power company installed its first burner, which consisted of a length of one-inch pipe flattened at one end into what was known as a "goosebill." Pauson injected steam into the pipe to make the oil flow freely. It took time, patience, and ingenuity to advance from the primitive burner to more sophisticated models, but the company persisted, and the oil-burning operation proved to be a success.

Since this operation was the first in San Francisco to burn oil as fuel, it created great interest among engineers, including those of the Market Street Railway and Matson Navigation. "The engineer for Matson," Pauson remembered, "stated that oil might be all right for use in San Francisco, but it could never be used on a boat!"

$100,000.00

CAN BE SAVED to the

PACIFIC COAST

IN ONE YEAR by all families
purchasing their

Illuminating Oils
IN BULK.

—

Oil in this Package 3 Cents
per gallon less than
in Cases.

Continental
Safety
Oil.

The Continental Oil and Transportation Co.

OF CALIFORNIA

Have introduced their TANK CARS for the transportation of Oils in
bulk from the refineries at the East, also the PATENT PORTABLE
BARREL (with reversible faucet) in which to ship bulk Oil to the
Dealers; and the 5 Gallon

DIV ME SUMFIN HARD

"LITTLE WILL" CAN

(which the cut on the left represents)

IN WHICH CONSUMERS CAN PURCHASE

CONTINENTAL SAFETY OIL

of their Grocer or Oil Dealer at 3 cents per
gallon less than the cased price for the same
grade Oil, and as over Four Million Gallons of
Oil are consumed annually on this Coast, the
above amount of money can be saved, if

Continental Safety Oil in **BULK** is used.   It is Water White, High
Fire Test, Free from Odor, and gives universal satisfaction.

It is sold in every City, Town and Village on the Coast.

Ask your Grocer or Oil Dealer for it, and take no other.

*Advertising petroleum products,
c.1880's. (California Historical Society
Library)*

The Emporium, Market Street, San Francisco; one of the first commercial buildings in the city to receive electricity from an oil-burning power plant. (Bancroft Library)

After proving the feasibility of burning oil, Pauson, like so many other enterprising men, returned to the San Joaquin Valley to acquire oil lands and wells to secure his company's supply. In boom-town Bakersfield, saloons, dance halls, and sporting houses ran wide open. Men frantically leased and re-leased land, fighting with fists and sometimes with guns to maintain their claims. Claims were staked and jumped by the dozens, and if one group of promoters went broke another moved right in to take its place. Kern River had become a major California oil field.

2 Cable Tool Days

The Chinese employed the cable tool drilling method at least 2,000 years ago to drill for brine from which they produced salt by evaporation. This method depends on the repeated dropping of a sharpened cutting tool to pound a hole into the earth. The Chinese "spring-pole" method depended upon human power to raise the tool. David and Joseph Ruffner used the same method of well drilling in 1806 near Charleston, West Virginia, in search of salt. Edwin L. Drake, "Colonel" by courtesy, used cable tools powered by a steam engine to drill for oil at Oil Creek in western Pennsylvania in 1859, bringing in a thirty-barrels-per-day well that marked the beginning of the oil industry.

In cable-tool drilling, a "string" of tools consisted of a heavy bit and stem on the end of a cable. As the bit pounded its way into the earth, it pulverized soil and rocks. At intervals, the string of tools was pulled out, water dumped into the hole, and the resulting "slurry" of drill cuttings removed by bailing. As the hole deepened, it was lined with steel casing to prevent it from caving in. Of cable tools, long-time California driller Clifford Davis said: "You dropped a heavy chisel hooked to a long cable and as you pulled up and dropped the chisel it broke up the rocks and clays. Then you had a bailer to bail out the pieces. The bailer was a long piece of pipe with a bail on one end like on a bucket. You fastened a cable on the bail and let it into the hole. On the other end of the pipe there was a flap or door that would open up when you dropped it in the hole. The cuttings and clay would push the door up or open and [the cuttings would] go up into the bailer. Then when you picked up on the bail the door would close, holding the clay and cuttings inside; you pulled it up to the top and dumped out the clay, rocks and water."

By the 1890's oil men had developed the cable tool rig known as the standard rig. A steam engine furnished power to

David Swartz and Hall Proudfoot, drillers imported from Pennsylvania in 1883 to work for Hardison & Stewart, forerunner of Union Oil. They are seen in the derrick house of a well at Tar Creek. (Union Oil Company of California)

turn a wheel (the band wheel), which was connected by means of a crank and a rod (the pitman) to a wooden beam (the walking beam) balanced near its center on a large wooden frame. The revolving motion of the band wheel caused the walking beam to go up and down like a teeter-totter, thus working the drilling cable attached to it. The drilling cable was wound up on a drum (the bull wheel). An iron bar (the stem) was attached to the end of the drilling cable, and to the stem was attached the drill bit.

The continual lifting and dropping of the bit pulverized strata at the bottom of the hole. After the driller had drilled a short distance, he pulled the bit from the hole and ran a bailer into the hole on another cable (the sand line), which was spooled on the sand reel. Since there was no clutch in the drive from engine to band wheel, in order to disconnect the walking beam when drilling stopped—for bailing or tool-pulling—the driller simply pulled a cotter pin and slipped the pitman off the band-wheel crank.

The drilling rope ran over a single pulley at the top (the crown block) of the wooden derrick that supported equipment lowered into the well. Since the rope ran over only one pulley, it could not develop any more pull than it had at the bull wheel. To handle the heavy work of running casing—the pipe that prevented the hole from caving—there was another spool (the calf wheel) with a line that ran over a block-and-tackle system. This line (casing line) made it possible to pull many times more weight than with the bull wheel and its single pulley.

A drilling crew consisted of the driller and his "tool dresser." The latter not only "dressed" the tools by sharpening the bits on a forge to the required diameter or gauge, but also took care of many other chores around the rig. "When I started in West Virginia," recalled Earl Delaney, a native of that state who migrated to the California oil fields in the early 1900's, "we worked two twelve-hour shifts, and a driller and tool dresser made up the crew. The drillers and tool dressers would dress their bits at the well, according to the kind of formation being drilled. A forge was erected in the derrick using coal, gas, or oil for heating the bits. Boiler fuel was the

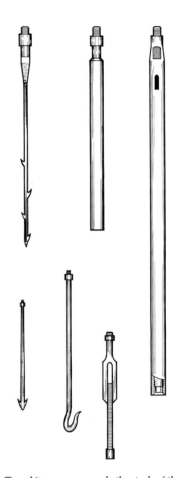

Turn bit, rope spear and other tools of the 1880's.

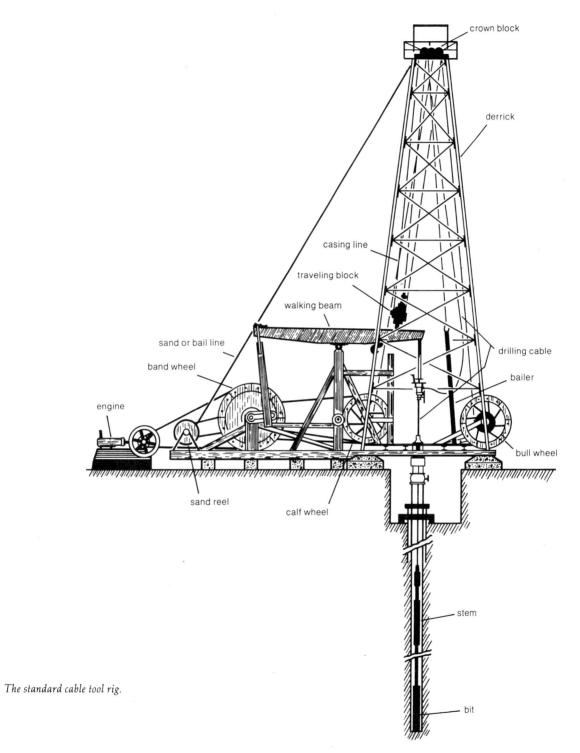

crown block

derrick

casing line

traveling block

walking beam

sand or bail line

band wheel

engine

drilling cable

bailer

bull wheel

sand reel

calf wheel

stem

bit

The standard cable tool rig.

23

Cable tool rig, Union Oil, Ventura County, c.1890. The walking beam that worked the drilling cable can be clearly seen at right center. (Union Oil Company of California)

same fuel as that used for the forge. When the $2\frac{1}{4}$ inch Manila drilling line became damaged, there was a trained man available to cut the line and put in what was called the nine strand splice. Usually it was such a neat job that you could hardly find where the splicing was done.

"Transportation was all by horse and mule power," Delaney explained. "You could hire a man with a horse and wagon for five dollars per day. Room and board was seventy-five cents per day. The drillers were paid $5.00 per day, and the tool dressers were paid $4.00. The company for which the drilling was done furnished the derrick, boilers, and engines. The contractor doing the drilling furnished tools, cordage, and personnel. In this area, the contractor would get $1.00 per foot, and in a wet hole area, they'd get a dime extra or $1.10 per foot. Each drilling contractor would have one drilling foreman or more, according to the number of wells he was drilling."

When the drilling crew lost a tool down the hole, it was necessary to recover the "fish." The first device developed for catching hold of a lost tool was a lock socket which would pass downward over the fish but would catch on any connection when raised. The socket worked well in shallow holes, but when wells became deeper the sockets (also called overshots) had to trade the piece that gripped and held the lost tool for slips—two or more pieces of wedge-shaped design—that would catch on the body of the fish and not merely on connections.

Jack Reed, who at age eighteen first worked for Union Oil in 1904 as a delivery wagon driver, was dressing tools for Sappho Enoch on one of Union Oil's Newlove wells in the Orcutt field when Enoch diagnosed a problem as a fishing job. "We came on tour early," Reed recalled. (In the oil fields, the working shift is called a "tour," pronounced to rhyme with "hour.") "I had to fire up the boiler and get it ready to run by 4 a.m. when Enoch was supposed to come on tour. When he came on, the first thing he did was let the pipe down, and then he ran the bailer to clean the hole out of any settlings that were in the well.

"When he pulled out the bailer, he went to sleep sitting over on the forge. I heard a lot of sand line rolling down on top of the belthouse." (The sand line was the line used to lower and raise the bailer. The belthouse was the enclosed portion of the rig immediately back of the derrick floor.) "I was out in the engine house and went in the rig and saw he had pulled that bailer into the crown [at the top of the derrick]. We looked up, but it was difficult to see without lights.

"He said, 'Kid, that bailer is stuck right up there in that wooden crown and we'll just have to wait around here until it gets daylight and we'll go up and tie onto it and then we'll loosen them bolts in that wooden crown and get it out of there.'

"We waited until daylight, looked up into the top of the rig, and no bailer! We looked around awhile and then he said, 'You know, boy, that bailer has fallen right down that oil well.'

"So he had to order up fishing tools as we didn't have any fishing tools on the job. We had horses then to haul them around, and the tools didn't get up to the rig until after we'd gone off tour. So he told the next driller what the trouble was and said he would have to fish the bailer out. When we came back on tour at midnight, the guy said, 'We can't find that bailer. There isn't anything in that hole. We've been to the bottom'.

"Enoch said to me, 'We'll wait until daylight and we'll set this string of fishing tools back and pick up the drilling tools and run in there and see if we can't clean that out a little.'

"At daylight I got up on top of the beam to screw the rope socket onto these fishing tools and I looked down into the bottom of the canyon about thirty or forty yards from the rig, and there the bailer was—down at the bottom of the canyon. I told Enoch where it was.

"Enoch said, 'Just take the sand line and go down and tie on and we'll report fishing it out.'

"We went ahead and got it all back up in the rig again and then when the other guy came on tour he said, 'How'd you get them out?'

" 'Oh,' Enoch replied, 'we just run the tools in there and stirred her up a little bit and got her back into the hole and went right in and fished her out.' "

Wooden derricks served to support drilling tools until the 1920's, when steel derricks began to replace them. The legs of a wooden derrick might consist of four tree trunks, hewed square. Rig-builders, the men who built the derricks, braced together pairs of legs on the ground, started the iron nails that would hold the legs together, and hoisted them by block and tackle to men in the derrick, who took great pride in being able to drive the nails home with one blow.

Jack Reed recalled a wooden derrick that burned. "The rig Sappho Enoch was drilling on burned down on a Saturday night." Enoch was not at work when the fire occurred. "He was supposed to go to work at 4 a.m., but didn't do it. He was

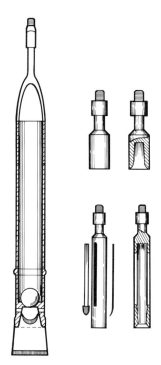

Cable tool crews ran casing as they went along to protect the walls of the hole. Hauling the necessary casing to the well sometimes became a family affair. (Kern County Museum)

Bailer (left) and sockets used in fishing jobs.

in the boardinghouse where we ate every morning and he heard his phone ring for the rig—two longs and a short. He got up and answered the phone, pretending he was on the rig. It was the drilling foreman, Lou Teatsworth. He asked Sappho how everything was going. Sappho said, 'Well, we just got cleaned out to bottom, Lou, and everything is in good shape.'

"Lou asked if there was anything he needed and Sappho ordered some little thing, I don't remember what it was and we could hear only Sappho's conversation. We all knew the rig had burned, but Sappho didn't. So he went up there to go to work on the rig, and the rig was all burned. He went up to another rig and called Lou up and said, 'I guess I'll have to countermand that order I gave you this morning. The damn rig is all burned down.' "

Tools and techniques changed to meet the challenge of drilling deeper holes. Wire rope began to replace hemp or "rag line." Wire rope had existed before the oil industry began and was used at first to run casing with the calf wheel. However, steel lines could not match the stretching action of the hemp cable for drilling. When well depths became greater, the diameter of the hemp rope necessary for the greater loads became too large to be practical, and drillers shifted to wire line on the drilling bull wheel as well.

The steam engine that powered the rig grew larger. Drake in Pennsylvania used a wood-burning steam boiler similar to those used on steamboats. The engine sat beside the boiler and generated about six horsepower, adequate for Drake's $69\frac{1}{2}$ foot well. By the turn of the century, manufacturers increased boiler output to forty-horsepower and engine capability went up to twenty-five or more horsepower. Producers also learned that natural gas could be used as fuel, and when it was conveniently available, natural gas fueled boilers for drilling rigs.

Steam furnished flexible, smooth, trouble-free drive. Steam engines started easily under full load, and their few working parts minimized maintenance problems. A valve easily controlled the engine. Yet problems existed that led eventually to the replacement of steam. Water could scale boilers badly and put them out of commission. At remote drilling jobs, there might be no gas, oil, or wood; fuel had to be hauled long distances.

At Drake's well in Pennsylvania, the drilling methods were those used for brine wells with one notable exception. Men drilling for brine set conductor (the first string of pipe in the well) by hand-digging a hole down to bedrock; conductor pipe was made of wood. Drake found surface sediments too thick and full of water to hand-dig a hole. He tried driving cast-iron pipe of four- or five-inch diameter, holding joints together by wrought-iron bands, the first known use of iron or steel pipe in a well. Driving the pipe proved successful, but there was not enough iron pipe available to replace the shallow wooden conductors immediately.

In a later development, drillers used "stovepipe," made by folding over steel plates two or three feet in length and riveting the seams. Usually the casing string was made up of two columns of riveted pipe, one fitting tightly inside the other. To get a better bond, the drilling crew indented the pipe with a sledge. Because there were no screwed connections or collars

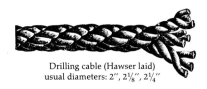

Drilling cable (Hawser laid)
usual diameters: 2", $2\frac{1}{8}$", $2\frac{1}{4}$"

Sand line (Hawser laid)
usual diameters: $\frac{7}{8}$", 1", $1\frac{1}{8}$"

Bull rope (Plain laid)
usual diameters: $2\frac{1}{8}$", $2\frac{1}{4}$", $2\frac{1}{2}$"

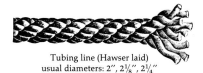

Tubing line (Hawser laid)
usual diameters: 2", $2\frac{1}{8}$", $2\frac{1}{4}$"

Sucker rod line (Hawser laid)
usual diameters: $1\frac{1}{2}$", $1\frac{3}{4}$"

Drilling cables like those shown (left) were used on rigs like the one (right) at Tar Creek in the Sespe field. The boiler in the foreground provided power for the rig. (Union Oil Company of California)

between joints of riveted pipe, the pipe was smooth on the outside and sturdy enough for driving.

Late in 1910, Standard Oil built a wooden derrick to drill a well on Section 14, T32S-R23E, immediately west of Taft. Company crews began to rig up on November 1 and, using cable tools, spudded the hole, No. 6, four days later. There was no shallow water, and crews, as was customary, made top hole without trouble, landing stove pipe at 545 feet. The stovepipe was sixteen inches in diameter. All went well as the pipe was set in the hole, but trouble followed.

The crew had only made a foot or two below the pipe when the bit became cocked over in the hole. Efforts to fish it out failed. One of the members of the drilling crew was John Stuck, a Pennsylvania Dutchman who was widely known as a meticulous and deliberate man and as versatile a hand as might be found in the oil fields. Among his characteristics was a unique habit of speech: he would interject "and so it did" in the middle of his sentences.

As the perplexed crew contemplated the fishing job that blocked them from going deeper, a solution presented itself —and so it did—in the wiry form of John Stuck. John was young and lean. Why not lower him down the hole to straighten the errant bit? John agreed to go. First, though, someone had the presence of mind to lower a flame into the depths to make certain there would be air for John to breathe. Fortunately, they had not cut any gas pockets. Having ascertained that there was sufficient air, Stuck removed his jacket and allowed the crew to tie a line under his arms. They lowered him down the hole, feet first.

There was no working room in the hole. John could hardly take a deep breath, much less bend down to work on the bit with his hands. He could, however, draw up his legs enough to kick. He kicked the bit until he had straightened it, then tugged the line to let the crew know he was ready to come up. Gingerly his fellow workmen pulled him out, not caring to lose another "fish" in the hole, particularly a human one. John Stuck reached the surface safely, put on his jacket and told the others he had figured it would take "some hard kicks—and so it did—to set the bit right."

Resourcefulness was a password among operators and drillers. "They simply had a lot of know-how," Leland K. Whittier said of his father and other early production men; "they could make out with little, and did! It was the ability to get a lot of work done with a minimum of tools to work with, because they just didn't have them."

Whittier's father was supposed to have once carried a walking beam on his shoulders. His son doubted that story, but remembered that "there are many tales about my father and Tom O'Donnell and their physical fitness. They were not

Workmen in Union Oil's Santa Paula shops built this steam engine in 1896 for use in the oil fields. (Union Oil Company of California)

exceptionally strong, but they had the know-how that gave them the ability to get things done. I remember one of their old friends said, 'You know anything that is round is made to roll. There are many ways of doing things, but it's just the way you go about it.' I don't think both of them were in the class of professional weight lifters or anything like that, but they just knew how to do things.''

Sometimes ingenious men literally invented tools on the spot to solve drilling problems. At Coalinga, R. C. (Carl) Baker, who had started his oil career as a twenty-year-old youth working in the Los Angeles City field, drilled twenty-four wells on the Inca property for Associated Oil. ''This West Side was a hard country to drill in,'' Baker told Frank Spurlock in 1957, ''because there were such hard streaks and then soft sand right underneath. They had to fish right at the bottom all the time," because after breaking through the hard streak the tools would go down too fast and, like a car skidding off the paved highway into sand, they would become stuck. "So I developed a side bit first, to drill the hole bigger than the casing," said Baker. The new tool worked well, but the driller always had to have three or four feet ahead of the casing, which "spoiled it" by making the hole too big. Undaunted, in 1905 Baker developed a "casing shoe" or device on the end of the pipe, with teeth on it, making it possible for the driller to work the casing through the hard places and keep it on the bottom all the time. "That was the start of Baker Oil Tool," he recalled; the Baker casing shoe became popular and in great demand.

To meet the needs of oil operators, machinists designed and built tools in shops that sprang up in Bakersfield and Los Angeles. Dick Smith recalled the beginnings of the Union Iron Works. Smith, who had been employed by Stiles & Parker Press Works in Middletown, Connecticut, at the age of fourteen, had served a five-year apprenticeship as machinist and toolmaker before coming to California in 1897 to work at the trade. He explained that around 1900 Union Oil grew dissatisfied with having all the repair work for its Santa Paula drilling operations done at a small shop in town. "Lyman Stewart, who was one of the co-founders of Union Oil, decided to build his own repair shop on the lease and made arrangements to get Ed Double to come out and take charge of the shop." Double had invented several important tools, and Stewart had known him in the Titusville, Pennsylvania, oil fields. Double "brought with him several highly skilled oil field machinists such as blacksmiths and joint turners." And so began Union Oil Tool.

The company grew so fast it was necessary to get a more central location. The firm moved to Los Angeles, and Smith went to work for it in 1902. "The shop in Los Angeles consisted of a machine and forge shop. In the machine shop there were two joint turning lathes, and a large or long lathe for making screen pipe for the completion of oil wells. The pipe was perforated and wrapped with wire on the outside to pre-

Carl Baker (left) stands next to a complete line of Baker casing shoes. With him is L. R. (Roy) McCollum, Baker's export representative at the time; Coalinga, 1914. (R. C. Baker Memorial Museum, Inc.)

vent sand from coming into the hole when the well was producing. In the forge shop were three forging hammers: one for jars, one for ordinary bits, and one for extremely large bits,'' Smith recalled.

Skilled machinists in Union's shop needed no drawings, Dick Smith observed. ''If a bit was needed, say an $8\frac{1}{4}$ inch with a four-by-five-inch pin, a little strip of paper [with the specifications] would be stuck onto a little sticker on the side of the wall. A man would take it off and go ahead and make it. He was so specialized in his work he needed no drawing.'' If a new man were hired, he, too, would make a bit without additional specifications. ''If you would take the two bits made by each of the men and lay them side by side, ''Smith said, ''they would be identical. The same thing was true of cable tool jars and other tools.''

Jars were devices used as part of the drill string to permit the tools to fall on the downstroke, but "jar"—give them a sharp jerk—on the upstroke which tended to loosen them from any crevices or cavings. The jars formed the connection between the sinker bar and the drilling cable and consisted of two links which slid on each other from six to thirty-six inches. Denny Driscoll, a talented Pennsylvania machinist who migrated to California in 1908 and earned a wide reputation as an ingenious tool designer, always insisted that nothing had contributed more to the development of drilling than jars. The old-time Pennsylvania blacksmiths had first introduced them in the boom days of Oil Creek, and Driscoll used to say that he had seen an occasional old-time steel-lined jar lying on the junk heap, worn out but never broken.

In the early days in the Los Angeles City field, wells tended to decline rapidly because producers did not know they should hold the gas in the ground to keep the oil flowing. After they had blown the gas out of the field, the wells produced only one or two barrels a day. The low production rates impelled many one- or two-well owners to sell their holdings, particularly since the oil had to be steamed by a set of boilers to below two percent basic sediment before it could be sold to the refinery. If a man did not have boilers and his neighbor did, the advantages of selling out became substantial.

Coincidentally, the invention of the Allen patented pumping unit spurred concentration of ownership. The first such units had made their appearance in the 1890's. The unit consisted of a vertical shaft driven by a bevel-gear. On the upper end of the shaft was an eccentric to which were attached the wires, or pitmans, from the various pumps, preferably in such manner that the pull of the pumps would balance one another. When this could not be done, a counter-balance was employed. The stroke of the pump corresponded to the revolution of the eccentric, being twelve to seventeen strokes a minute, according to the gravity of the oil and the amount of oil pumped at each stroke. With wells a great distance apart, wire cables made connections to a reciprocating jack. In some instances, this method conveyed power for more than one-half mile. Steam and, in a few instances, gas engines furnished

Sketch of a jar.

The Allen patented pumping unit made it possible to pump many shallow wells from one central power source, at Kern River (above) and other oil fields. (Kern County Museum).

The circle type pumping jack drawn at right linked each well to the Allen or "jackline" pumping unit.

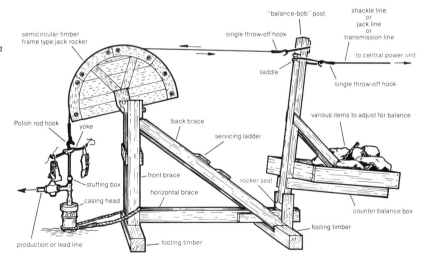

"balance-bob" post

shackle line or jack line or transmission line

single throw-off hook

to central power unit

saddle

single throw-off hook

semicircular timber frame type jack rocker

various items to adjust for balance

back brace

servicing ladder

Polish rod hook   yoke

rocker seat

stuffing box

front brace

counter balance box

casing head

horizontal brace

footing timber

production or lead line

footing timber

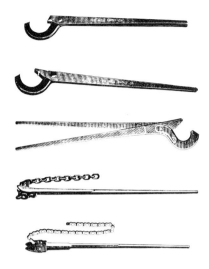

Casing tools used in cable tool era.
(*American Petroleum Institute*)

the motive power. Oil was the usual fuel, as natural gas was used only to a limited extent.

By 1905 the Allen patented pumping unit, soon nicknamed the "jack-line" system, saw wide use. "If you had a jack-line plant and a steaming plant, which is a dehydration plant, and the person adjacent to you had a well on a lot, this well could be purchased," said Kenneth Manley, whose grandfather and father were pioneer Los Angeles oil operators. "You merely ran a cable from your jack-line system over to the existing well and piped the oil into your tank and you were in business." It made more economic sense for the neighbor to sell his single well on a single lot than to go to the expense of installing his own jack-line and dehydration equipment.

Keeping completed wells on production posed problems. The shallow wells had a tendency to sand up, which curtailed and often completely stopped the flow of oil. To remedy this, the operator had to pull the tubing from the hole and clean out the sand. This arduous task is known as "pulling the well."

"One of the first ways to pull a well," Kenneth Manley recalled, "was to have a windlass with a ratchet on it over one leg of the derrick. The crew of two men had blocks and elevators—fairs, I believe they called them." The derricks stood about fifty-two feet high, and the men would wind on the windlass and pull up a "stand," or two joints of twenty-foot tubing. They had a set of slips that they worked with their feet which they used to lock the tubing in the hole. "Then they would relax on the windlass, take their crummy old tongs and unscrew the joint of tubing," Manley said. One of the men had to climb the derrick and set the stand back, and then they would stack it. This procedure meant that every time they pulled a stand of pipe, a man had to go up the derrick and down again.

One of the unusual events that occurred in pulling wells was related by Earl Delaney, who worked in the old Salt Lake field in Los Angeles shortly after the turn of the century before moving on to Taft and a forty-one-year career with Standard Oil. A crew was dispatched to pull a well, but found the steam line under water from heavy rains. No one could get the engine started, and the men left, leaving the engine in gear,

Tool dressers and machinists like these pictured at Santa Paula, c.1890's, built up a reputation as mechanical wizards who could redesign or modify almost any kind of oil field machinery. (Union Oil Company of California)

with its brake off. "Some time in the early morning hours the engine worked the water out to where it could run," said Delaney. "It pulled some two thousand feet of sucker rods and wound them on the bull-wheel shaft," Delaney said. The next morning the surprised gang had to cut the rods off the wheel with a cold cutter.

Well-pulling continued to be a hard task for a surprising length of time, Kenneth Manley recalled, until "someone in the Doheny organization" began experimenting with the use of horses for the work and initiated a great saving in labor. "A boy would lead a team of horses by hand as they pulled the tubing out of the hole," Manley said. "That way, one of the men in the production crew could work derrick, and one could unscrew the pipe, while the horses did most of the work. In fact old-timers claim the question often came up as to who was the smartest on the rig: the horses or the well-puller who was doing the wrenching. The horses, after a while, realized what their job was, and, as the story goes, they could not only pull a string of tubing out of the hole, they could put it back in, too, and they knew where to stop when the slips were set."

Pico Canyon at Newhall was the scene of some of the first commercially successful oil production in the state. Pico Canyon's appearance had not much changed from 1885, when this photograph was made, to 1917, when Ross McCollum went to work there for Standard Oil. (Title Insurance and Trust Co.)

As late as 1917 horsepower was more than a figurative term. After Ross McCollum graduated from the University of California at Berkeley, he went to work that year for Standard Oil at Pico Canyon, not far from McCollum's Los Angeles home. The company hired McCollum, whose degree was in chemical engineering, as a roustabout. He drew as an early assignment a well-pulling job on Pico No. 4, which had been completed with wood rods. McCollum had to lead a horse which pulled the rods out of the hole. "Of course, the horse knew more about the job than I did," he said "and at the proper time, by a slight wiggle of his rear end, would either pull the pin out of the socket, or the tubing out of the collar, or lower the rods into the socket and the tubing into the collar when going into the hole." This kept happening and all McCollum can remember is "the size of the horse's hoofs and how close he came to stepping on my feet—although of course this never happened. Every day that I worked on this job the horse's hoofs looked larger."

In every oil field, there were those who thought they had the answers to production problems. In 1901, J. W. Goff, of

San Diego, visited in Bakersfield and, while cheerfully acknowledging that he did not know anything about the oil business, nevertheless claimed to have an answer to a problem that had begun to bother some operators in the Kern River field: the existence of low-gravity,* tar-like oil, particularly on Section Eight on the south side of Kern River, that was all but impossible to produce.

Goff obtained permission to try an experiment on a well on the Golden Rod property. He ran a steam line into the well to heat the oil in the hope of making it more fluid and hence easier to produce. When steam did not achieve the desired result, he ran air into the hole, also unsuccessfully. Then he rigged the air line to run through the steam line so that steam would heat the air, and in this manner hot air was injected into the well. The well began to flow steadily for a week, but with the price of oil declining, the cost of the operation exceeded the benefits. Goff disappeared, a pioneer in using heat in a way that would eventually revolutionize the field.

Resourcefulness did not remain confined to technological matters. It was not uncommon, for example, for a crew to "hold out" the amount of hole they had drilled in a single day, reporting a depth considerably shallower than what had actually been reached. The difference afforded them leeway if they encountered hard drilling and gave the appearance, at least on reports, of their making steady progress.

At Union's Graciosa No. 15, Jack Reed and his drilling partner took over from a crew that had lost a string of tools down the hole. Before the first crew left the job, they managed to drive casing by everything but the bit, which posed a difficult obstacle. The foreman said to drill an open hole, that is, a hole unprotected by pipe, below the casing until they picked up oil sand. If they did not find sand, the foreman said, they would not bother to run casing deeper. The going was easy,

---

* Gravity is a measure of the density of liquid petroleum. The measurement is in terms of comparison with water. Petroleum liquids lighter than water have gravities greater than 10°, on the scale established by the American Petroleum Institute. Those in the general range of 30° and upward are often called light or high-gravity oils; those in the range from 10° to 15° are often called low-gravity or heavy oils.

Children sometimes accompanied their dads during work at the wells. This scene was photographed at Robertson No. 1 at Bardsdale in 1890. (Union Oil Company of California)

but Reed and his partner did not turn in more than twenty-five feet of hole a day.

"One morning," Reed recalled, "the drilling foreman said to me, 'Exactly twenty-five feet every day! Don't seem to make a damn bit of difference. You can't make twenty-five feet every day. Why don't you turn in a foot more or less? How much hole you got held back?'

" 'Oh,' I said, 'we haven't any.' "

When they hit oil sand, the sand was two hundred and fifty feet deeper than the depth reported for the well.

" 'Holding back two hundred and fifty feet of hole,' the foreman said. 'You are gonna get me in a mess doing this kind of job.' He continued, 'Well, Jack, we're going to have a hell of a time getting that pipe by the bit down there.'

"We did, too. It took us about four or five days to get by it. 'Well,' I said to the foreman, 'you won't have a lot of alibis as to why we aren't getting by the bit. We just held this hole back to make it look good for the whole bunch of us.'

"He said, 'That's too damn much hole.'

"But we went ahead and ran the reamer to bottom and ran the pipe and then finally drilled the oil sand and set a liner and brought the well in."

3 Organizing Companies

Sometimes the people who organized companies made themselves and others wealthy. Sometimes they left their investors poorer. At times, in their search for the financial backing to drill wells, they involved celebrities in their schemes. At all times, they played a significant role in the beginnings of the oil industry in California.

W. R. Guiberson was manager of Associated Supply at Coalinga in 1909 when he joined with others to organize the Silver Tip Oil Company, of which he became president. The role of company president was new for Guiberson, who had begun his oil career as a roustabout and tool dresser for Union Oil at Torrey Canyon, later becoming superintendent of Commercial Petroleum at Coalinga, and still later, manager of Bunting Iron Works before joining Associated Supply.

Guiberson and others incorporated Silver Tip Oil for $75,000 with 75,000 shares of one dollar par value. Among them, Guiberson, Z. L. Phelps and H. A. Whitley, all of Coalinga, and Dr. E. R. Smith of Los Angeles, held more than 45,000 shares and, with the shares, controlling interest. They sold additional shares in Coalinga. With the proceeds, they purchased twenty acres on Section Six, paying $1,500 an acre, or a total of $30,000. Silver Tip Oil spudded its first well soon afterward.

Like many another oil venture, the Silver Tip had its less encouraging moments. A contractor who did teaming work for the company in April, 1909, agreed to accept payment amounting to $500 in the form of stock shares valued for purposes of the payment at 75¢ a share. When it came time to collect, the contractor found stock trading for 66⅔¢ a share.

Operators and their families; Grand Oil Company well; East Sunset, Kern County, January, 1904. (Kern County Museum)

Coalinga Oil Record

VOL. VI. NO. 31. COALINGA, FRESNO COUNTY, CALIFORNIA, SATURDAY, SEPTEMBER 25, 1909. $3 A YEAR.

SILVER TIP PROVES GREATEST GUSHER EVER DEVELOPED IN THE STATE

BLUE MOON WELL ON THE PUMP SHOWS A THOUSAND BARRELS

The Silver Tip well the morning of September 22, 1909. The bailer which was blown out of the hole, is seen between the girths on the right side of the derrick. Superintendent Phelps is in the foreground, wearing a white shirt, his hands grasping his suspenders and with a smile on his face.

FLOWS ALMOST UNPRECEDENTED STREAM FOR TWENTY-FOUR HOURS

One of the greatest occurrences in the history of the oil industry of the West happened in Coalinga Wednesday and Thursday of this week when the Silver Tip well on section 6 gushed a steady stream of oil, for more than twenty-four hours, estimated by oilmen experienced in the coast and eastern oil fields, all the way from 10,000 to 22,000 barrels.

The well began flowing Tuesday evening, but a bailer that had been lost early in the day became caught near the mouth of the pipe and not much oil forced its way through during the night. The following morning, Wednesday, at about 7 o'clock, the bailer was blown out and the well started spouting oil at a tremendous rate. Instead of subsiding, the flow increased as time wore on and by evening the increase was very perceptible. It continued on into the night, emitting much gas at the same time. A worthy feature was that during the day a very small amount of sand was thrown out, but toward evening this increased and while the night, the sand increased also until finally Thursday morning, shortly after 8 o'clock, the pipe choked up. Sand is heaped up on all sides of the derrick and in the belt house it is several feet deep. Much work will be required to clean away the sand

5:30 a. m., W. A. Hersey, civil engineer and licensed surveyor, made a measurement in the ditch leading the oil from the gusher to the reservoir and at that hour it was flowing at the rate of 18,140 barrels a day according to his figures. A half hour later he made another measurement and the rate had increased to 21,160 barrels, and this after a steady flow of twenty-three hours. Mr. Hersey declares that his estimates are very conservative, as he deducted thirty per cent from his actual figures to cover all probable discrepancies. Yesterday afternoon Colonel Fred E. Windsor, who for years was an oil scout in Pennsylvania, made a measurement. He estimated it at 720 barrels an hour or 17,280 barrels a day. Colonel Windsor made a liberal deduction from his actual figures. A well known west side engineer made a very conservative estimate, after a deduction, of 600 barrels an hour or 14,200 barrels for the day.

A Great Spectacle.

The spectacle was a treat of the rarest kind and from early morning untill dusk a steady stream of automobiles and other vehicles traveled to section six. Quite a wind was blowing in the morning and the spray was heavy. The ground for many acres around was saturated. Throughout the field and in town the news

PRODUCERS WILL BE PUMPING OIL BY THE FIRST OF THE YEAR

Rapid progress has been made in the construction of the pipe line of the Producers Transportation Company during the past few weeks, and according to the statements of the officials of the line, oil will be pumped over the mountains by the first of the year. The pipe will be laid by that date and the line practically completed with the exception of the houses at the different pump stations along the route.

About sixty miles of pipe have been laid to date. Tuesday of this week the tong gang, working from this end, had reached the fork in the Dudley road and by the end of the week Station 2, about thirty miles south, was reached. Pipe is being laid at the rate of a mile a day. Almost fifty men comprise the gang. About equal progress has been made from San Luis Obispo.

Another Tank Filled.

At the tank farm south of town, the second 55,000 barrel storage tank has been filled with oil, the petroleum coming from one of the big American Petroleum tanks. Monday the third tank will be completed and in about a week the fourth. This number will be all to be erected at the farm for the present. Two will be erected at Coalinga Station 2 next and two at Coalinga Station 3. Men are at work extending the feed line

About forty miles of line have been built out of San Luis Obispo.

A camp for tong gang No. 3 in the south has been installed at McKittrick, at which end L. H. Cory is the district foreman in charge of the operations. Pipe laying will begin at this end of the line about the middle of next week. This gang will operate between McKittrick and the Junction on Antelope plains. Commissary E. H. Gould has been in McKittrick and Bakersfield for several weeks past establishing the camps.

At the Stations.

Last week three 14 ton Heine boilers for Coalinga Station 1 arrived from the east. They are the largest single pieces ever hauled out of this field. Cheney Brothers did the teaming and put twenty-four horses on each boiler. Three boilers of like make and like capacity have arrived at Huron for Station No. 2. Robert Forst of San Francisco has secured the contract for the cement and concrete work on the pump stations and will begin next week. Engineer R. J. Reed has just returned after staking out the locations for the buildings of Coalinga Stations 1, 2 and 3. The pumps, which will be furnished by the Fairbanks-Morse Company, are due to arrive from the east the first week in September. They are monster affairs, when set

Banner headlines greeted the Silver Tip gusher. (Coalinga District Library) Shares of company stock, issued at $1 par value, quickly soared to $4 a share after the splendid show of oil. (R. C. Baker Memorial Museum, Inc.)

44   *Organizing Companies*

On September 4, 1909, the headline across the top of the *Coalinga Oil Record* proclaimed, "Silver Tip on Section Six Gets Light Oil." The newspaper reported that the well had broken loose. For four hours it intermittently spouted a stream of oil and sand that tore away one side of the derrick house, broke the main crown pulley, splintered the crown blocks, and "wasted oil all over the adjacent ground." Observers reported that rocks and hard balls of sand striking against iron parts of the drilling outfit caused flashes of fire, but wind blew out the flames without additional danger. Gravity of the oil was estimated at 23° or 24°, indicating higher quality than much of the oil previously produced in the Coalinga field. Although the well sanded up a week later, the value of Silver Tip shares reflected the strong showing of oil. An owner advertised 300 shares for sale at $2.25 per share.

Out at the Silver Tip well, the drilling crew ran a bailer into the hole to clean out sand and mud. The bailer became stuck. The well began to flow, first around the bailer and then, after blowing the bailer out of the ground, through the well bore, spraying an estimated 10,000 to 20,000 barrels a day. Motorists drove out to see the display, hailing the well as the greatest gusher ever seen in California. Workmen rushed to build sumps in nearby gullies to hold the oil. After a week, the flow continued at the rate of an estimated 10,000 barrels per day, making a roar clearly heard in Coalinga, a mile and a half away. A newspaper reporter, kept awake by the roar in his hotel room, described the sound as "like a score of locomotives blowing off steam." The spray of oil carried half a mile, and in one seventy-two-hour period, the flow to tanks amounted to 36,000 barrels of oil.

In October, hardly more than a month after the first show of oil, Silver Tip directors declared a ten-cent dividend, the first of what would be regular monthly dividends. With oil bringing forty to fifty cents a barrel, the well produced enough during its first month to pay drilling costs of $27,000 and provide a surplus of $24,000.

Thanks largely to Silver Tip, the town of Coalinga took on new confidence. The local newspaper jauntily proclaimed on its banner: "Coalinga Will Have 10,000 Population in Two

Years" and "Read Oil Record, Then Mail To Your Friends in East." Two representatives of the Coalinga Athletic Club prepared to go east themselves with an offer of $10,000 to boxing promoters looking for a site to stage the long-awaited Jim Jeffries—Jack Johnson heavyweight title fight. Such boosterism was common to new oil towns, and enterprising businessmen made the most of the civic pride and promotional energy that went along so well with their own efforts.

Financiers took notice of the Coalinga oil development; the Sierra Madre Club, a mining and oil promotional group from Los Angeles, scheduled an excursion to bring "bank presidents, men of capital, and leaders of industry" to tour the Coalinga field. The Los Angeles Stock Exchange closed on the

Long lines of automobiles waited at the Coalinga depot to meet the Los Angeles businessmen's special. The visitors were greeted by a large number of townspeople as they stepped from the train. (R. C. Baker Memorial Museum, Inc.)

Friday the tour left on a Southern Pacific special train, because most of the exchange's members had become oil field tourists. "All aboard for the big Sierra Madre excursion to the oil fields," proclaimed a sign on the exchange floor.

On a November day, the special twelve-car train bearing 200 excursionists arrived in Coalinga to be greeted by the town's citizens and a band, which heralded the arrival with "Jungle Town." Along with a tour of the oil fields, where the Silver Tip well continued to flow, the visitors' stay featured a Saturday night banquet at the Arcade Rink. The menu included "oyster cocktail, ripe olives, celery, cream of tomato soup aux Crouton, French crab meat en Cocile, asparagus Mayonnaise, roast mallard duck, sweet potatoes glace, sparkling Burgundy, Neopolitan ice cream, assorted cake, cafe noir and cigars." Members of the Ladies Improvement Club served as waitresses. Captain J. F. Lucey of Lucey Supply Company was toastmaster, and the Derrick Quartette sang. The quartet's members were Dr. C. H. Warren, Thomas H. Pike, Roy McCollum, and W. R. Guiberson. Two weeks later, Silver Tip's directors declared another ten-cent dividend.

Silver Tip Oil was one of many companies formed in the Coalinga field. J. W. Pauson, who had brought in his first well several years before at Coalinga, stayed on in the field after that initial success and participated in the formation or ownership of various companies, including Kaweah Oil, Twenty Six Oil, Tavern Oil, Lorene Oil, Independence Oil, W. K. Oil, Turner Oil, Coalinga Mohawk, Arline Oil, and Wabash Oil. No company would be more famous than another company in whose beginnings he was involved in 1910.

"Captain John Barneson, Joseph Seeley and Tommy Turner were interested with me in the Techau Tavern Restaurant in San Francisco," Pauson recalled.

Barneson, a former Navy captain, had a few years earlier in cooperation with William Matson of Matson Navigation Company conducted experiments using oil as a marine fuel, successfully converting the S.S. *Enterprise* from coal to oil and completing several trial runs between the mainland and Hawaii.

"I needed additional capital and associates for my oil in-

terests," Pauson said, "and asked them to join me, which they did. From this chance beginning, Captain Barneson started General Petroleum." The company later became an important part of Mobil Oil.

Not everyone who invested in an oil company became rich, nor did they even get a run for their money. The Silver Tip well had hardly begun to flow before a full-page ad in the *Coalinga Oil Record* announced formation of Coalinga Crystal Oil to drill on Section Twelve. The company offered 20,000 shares at twenty-five cents a share. Two months later, the newspaper carried this headline: "Promoters and Oil Companies Under Fire." The accompanying article said that stockholders had found "mismanagement and frenzied finance in the conduct of affairs" of various companies, among them Coalinga Crystal.

Wagon tracks lead into the headquarters of San Joaquin and Kern Oil; these outfits joined other Kern River independents in forming Associated Oil Company in 1901. The woman in the foreground is unidentified. (Kern County Museum)

To promote sale of stock, various companies used the United States mail extensively, sending out glowing flyers telling of valuable properties and plans to develop them. In June, 1911, Postmaster General Hitchcock issued a mail fraud order against Haiwee Pacific Oil of Oakland, charging the company with sending a "fraudulently exaggerated" prospectus through the mail. The trouble was the latest in a long series for the Oakland company. The firm had started life three years before as Roosevelt Oil, not only borrowing the name of the then-president of the United States but also using his picture in its promotions. When the president ordered an investigation, Roosevelt Oil became Haiwee Pacific. Soon afterwards, a government investigator, stating that it cost no more than $4,000 to drill a well, charged the company with collecting $150,000 from its subscribers, of which $55,000 was used in one year for traveling expenses of the company's secretary and for other purposes not directly related to drilling a well. A judge found the secretary guilty and sent him to jail for six months and fined him $1,000. On appeal, the secretary gained his liberty, and the postmaster general said in his mail fraud order that the company was continuing its activity.

At about the same time in Los Angeles, police initiated a search for three men who had organized Saxton Oil. Mrs. Maud Taylor, a widow, charged that the three had duped her of her whole fortune of $40,000. Mrs. Taylor said she had bought stock thinking the company a good buy. When she became suspicious, she attempted to locate the men and was unable to do so. When police investigated, they discovered the company had been capitalized for $100,000 but that records listed only $25 worth of stock as having been sold.

One of the tasks that faced any company was the design of an adequate stock certificate. In at least one instance in the early days of the Kern River field, the task posed a problem. Henry Ach, president of Monte Cristo Oil, one of the field's earliest and most successful producers, decided in 1901 to improve his company's stock certificate with a new design. He told the artist that the center of the certificate should depict the Count of Monte Cristo throwing up his hands in amazement, while in one corner there was to be a drawing of the Chateau d'Iff where the Count was imprisoned. When the design came back, in one corner there was a burglar with a mask, dark lantern and whiskers. "What is it?" Ach asked. "What you ordered," the artist reportedly replied. "The Shadow of a Thief."

From time to time financiers talked of forming a stock exchange in Bakersfield. Stockbrokers took ads in the newspapers urging that people get on the bandwagon. After each new strike or well-publicized gusher, new companies —dozens each week in boom times—sprang into being as people tried to cash in on the new discoveries. No company, however, had quite the character of one formed in Los Angeles during the peak of excitement over the Kern River field.

In *Petroleum in California*, published in 1901, Lionel V. Redpath wrote: "In this progressive age when women are invading every branch of business and every profession and where their sphere of influence is as broad as the Universe, it is not remarkable that we should have a women's oil company. Neither is it strange that the company should be a success."

Women's Pacific Coast Oil was organized by women who barred men from any role but buying stock. The company, according to Redpath, owned 320 acres in Kern County, lots in the Nob Hill tract, Los Angeles property at "Placeritos Canyon" near Newhall, and a lease on lands at Summerland where a well was being drilled. "Work on the first well is progressing nicely," Redpath wrote, "and as the boring is comparatively easy, it will not be long before the company expects to have a producing well."

Women's Pacific Coast Oil was capitalized for $300,000, of which $100,000 was in the treasury, according to Redpath. Par value of the stock was $1 a share.

"From the beginning of the organization," Redpath wrote, "Women's Pacific Coast Oil has been a success. It was a success because its affairs have been managed in the most conservative and business-like manner and because the promoters were shrewd enough to secure the cream of the land in the district that they proposed to operate. It was a success because the public was satisfied, after an investigation, that

(*Bancroft Library*)

Investment brokers placed magazine ads to attract potential shareholders. The Los Angeles firm (top) offered a folder on "Money Making in the Land of Palms."

the officers, conscientious, painstaking, honorable individuals would handle the affairs of the company with the same care that they would their own personal business. In these days of fake oil companies Women's Pacific Coast Oil stands like a beacon light, pointing the way to legitimate success in oil enterprise."

One of the first tenets of raising money to drill a well was that the promoter had to go to people who had money in the first place.

While spending the winter of 1910–1911 in California, Frank (Husk) Chance of the Chicago Cubs decided to take a flyer in oil and invested in lands at Lost Hills about forty miles northwest of Bakersfield.

In the spring of 1911, a newspaper reporter in New York wrote of the Chicago Cubs baseball player: "Chance is smiling even in defeat. While his downtrodden athletes were feebly fighting against the irresistible Giants this afternoon, the peerless leader received a message from the West telling him of a rich strike in oil.

"Sometime ago Frank and his wife, who is making a western trip, invested a neat chunk of money in oil stocks. The well in which Frank's company placed reliance is located near Bakersfield, Cal. All told the Chances' investment amounts to about $13,500 face value.

"The stock has gone soaring with tremendous strides, according to a message received by Chance, which says that a rich oil stream has been found, from which the promoters expect to reap a golden harvest. Tonight the baseball wonder is painting a picture of great affluence for the message tells him, 'It's worth several millions.' "

Chance of the Chicago Cubs was not the first, nor would he be the last, of the celebrities who would be brought to the oil fields. In June, 1923, Sir Arthur Conan Doyle, the creator of Sherlock Holmes, showed up at General Petroleum's property at Ventura to look over the field. The English author, acquainted with high officials of Shell Oil in London, posed by a tank with General Petroleum's production manager, W. L. Philbrick, and later visited properties of Shell and Associated Oil. Whether he invested any money is unknown.

Railroads appealed to the interest in oil investment. This Santa Fe Railway ad also emphasized the line's use of oil as fuel. "No cinders, no smoke."

Companies provided housing for personnel. Frank Foster, gardener at California Oilfields, Ltd.'s camp near Coalinga, took this photograph, c. 1911. (R. C. Baker Memorial Museum, Inc.)

The boarding house for drilling crews was a featured element in Grace Oil's ad. (Beale Memorial Library, Bakersfield)

53

Promoters did all sorts of things to raise money. Garth Young, a lab technician and plant foreman at Signal Hill who later became senior vice-president of Signal Oil and Gas, recalled the boom conditions at Signal Hill in 1921. His boss, Sam Mosher, was a "super salesman" who kept the operation supplied with money for expansion. "He would show up anytime day or night with an interested prospect and, using a stopwatch, he would measure the production rate into a five-gallon can, filling it from the 'look box' near the storage tanks. He would calculate the twenty-four-hour production from the number of minutes it took to fill the can. It was an impressive way to raise money, and he never failed. Many of the early investors held onto their stock and lived to see it multiply more than a hundredfold. They always thanked him for the chance to buy in and help measure the gasoline production by the five-gallon can method," Young said.

At Signal Hill, burning flares from the natural gas released in drilling operations fostered a carnival atmosphere. The field became a promoter's paradise. Large red and white circus tents were set up in which busloads of prospective investors were fed meals, and even the waiters were high-powered salesmen. Driller Clifford Davis, who recalled the boom days vividly, said that he and his crewmates would "visit some of the sucker tents where they was selling stock and units in a well. They would get a barker outside of the tent and a spieler inside, feed a barbecue, then start their spiel. There were many of these around: McAdoo, Parkford, Foster, McIntyre Brothers, Meltzer, Santa Fe Ball, Toddie Boy, and the big one was C. C. Julian. He was king of all of them. He made Saturday and Sunday his big play. Before the spiel got through, the people was flocking to the counter table to shell out their $100 bills. The people got receipts for their money," Davis said, "but when they got their stock certificates it would be in some other well or just wherever the jugglers wanted to put the money."

Promoters also had a field day in leasing operations. After any new oil discovery, there ensued a scramble to lease land in

Incorporation papers for Union Oil Company were signed in 1890 in the second-floor corner office of this building in Santa Paula. California Oil Museum now occupies the first floor. (Union Oil Company of California)

Shasta Oil Company well, Kern River, 1901. (Kern County Museum)

the area upon which wells could be drilled. Each company's agents competed to gain exclusive rights to promising locales. The laws were complex, the opportunities for sharp practice were great, and stories abound of the opportunities missed as well as the deals made. During the boom at Santa Fe Springs in 1922, an agent for Amalgamated Oil, Joe Jensen, found three property owners who were willing to combine their small lots into a "community lease" by which a lot large enough to drill on could be created. All three owners would have a share in the production royalties. A rival landman offered the three owners $2,000 an acre and a one-sixth royalty, but said if they left the room the offer would be withdrawn. They indeed left the room, having promised Jensen the first opportunity to acquire the lease. Jensen's superiors, however, turned the lease down, saying that they had enough land leased at that moment. The property eventually was leased to Shell Oil, and that company produced millions of barrels of oil from it.

On another occasion, Joe Jensen recalled, he knew a man who had leased 100 acres to a promoter, who then placed a $100,000 cash bond in the bank as assurance of his intent to begin drilling a well promptly. When he failed to drill the

well, the landowner impounded the money in the bank, demanding that the well be drilled. Realizing that the land lay outside the area of productive drilling, Jensen advised the landowner to take the $100,000 out of the bank before the promoter withdrew it and backed out of the deal. The landowner followed Jensen's advice and later told him that he was a nice young fellow and in appreciation of Jensen's suggestion he and his wife were giving him $500. This for advice, Jensen recalled, that had netted the landowner one hundred thousand dollars.

Differences of opinion over the value of wildcat land were, of course, nothing new. Sometimes they played a significant role in the fortunes of a particular company. In 1928, Milham Exploration discovered oil at Kettleman Hills. In the aftermath of the big discovery, various companies jumped in to lease land, but, as usual, there was no agreement as to which lands might prove to be oil-bearing.

One of the companies that interested itself in Kettleman Hills was Superior Oil, formed by William Keck, a thoroughly experienced driller. He was one of the first to use rotary tools in California and earned additional fame in the Coalinga field as the developer of the "swinging spider," a device that helped with the running of casing.

When Keck learned that a particular property, the Huffman lease, was open, he looked it over and decided the land had promise. That at least two other companies enjoying success at Kettleman Hills thought the property worthless did not deter him. Keck, with the backing of geologist Walter English, decided to lease the Huffman property and completed the first well as a big producer in 1930. This development gave Superior Oil a big boost, and Keck decided to expand. One of the first men he hired was Bob Hutcheson, a paleontologist, one of the first such experts to be hired by the independent producers. Hutcheson recalled that although Keck had a reputation for being "rough and tough," he was actually "very reserved. But I could always talk with him." Hutcheson also recalled that Superior became known for drilling deep wells. Keck believed, said Hutcheson, that "when you started a well you should keep going until you knew it was hopeless."

Los Alamos Oil & Development Company's Logan No. 1 was the deepest producing oil well in the world (4,300 feet) when geologist Ralph Arnold took this photograph in 1906. (Huntington Library)

Deeper wells meant new opportunities in the oil fields. Sometimes production people in the field organized new companies to capitalize upon opportunities they came across. In 1922, for example, John E. (Brick) Elliott, an experienced geologist attempting to establish his own company to take samples of earth, known as "cores," from drilling wells, had a chance to lease land and try some wells of his own. He organized a new company, Elliott Petroleum, and after much hard work, his wells became excellent producers. Within two weeks after the first well came in, Elliott's partnership cleared itself of debt, and for the next several months offered its stockholders a one hundred percent dividend which then dropped to sixty percent for the remainder of 1923. Other promoters attempting to sell stock to the public on areas as small as thirty by one hundred feet often used Elliott Petroleum's dividend policy to convince potential investors of the quick road to riches that lay ahead.

Even as oil companies grew, tool companies expanded. Carl Baker, who had invented the Baker casing shoe in the Coalinga oil field, found Coalinga too small a base from which to operate the growing business of Baker Casing Shoe. He purchased property on Slauson Avenue in Los Angeles and built a plant. "The thing just kept growing," he recalled. "It kept getting bigger and bigger." Baker took charge of both the San Joaquin Valley and Los Angeles ends of the growing business. "I was down in Los Angeles a week and up in Coalinga a week, driving over the old Ridge Route." The trip was an arduous one. "I'd leave Coalinga at three o'clock in the morning and get into Los Angeles at two o'clock in the afternoon." In the spring of 1922, Ted Sutter went to work as office manager at Baker's new plant. "I was the one and only office employee," Sutter recalled. "We had a machine shop foreman and four or five machinists, most of whom had been transferred from the Coalinga shop, which had about five employees."

Later Baker built a $1.25 million plant on East Slauson Avenue, and still later built another plant in Houston, Texas. The business grew until Baker Oil Tools had over 1,200 employees and a payroll of more than half a million dollars a month.

"I had ninety-five cents when I started in for myself," Baker recalled, "but I always stood on my own two feet. I never did buy anything unless I had the money to pay for it. When our income tax began to get pretty heavy, we put in a system so that we knew every month about what our income tax was. And I told the boys, 'Now, that money doesn't belong to us. Set it aside.' Every month we put a certain amount of money for tax purposes in an entirely separate bank. You'd never catch us unprepared."

Of those who made fortunes from the oil industry, there

were even some who realized great returns without meaning to do so, through a quirk of nature and a combination of circumstances.

In pre-World War I days, a promoter bought a few acres on the edge of what was then an unpopulated area at Huntington Beach. The promoter intended to subdivide the acres into twenty-five foot townlots and sell them for a profit to a public eager to buy townlots almost anywhere in Southern California. The boom faded before the promoter could place his lots on the market.

About the same time, a printer in New England invested his money in a set of plates for a then-new encyclopedia. He printed and had bound several thousand copies of the set, but found the public reluctant to buy.

A mutual friend brought the printer and the townlot promoter together with the result that soon afterward New England newspapers carried an advertisement informing readers that for a small sum down and an agreement to make further payments, they could obtain not only a modern encyclopedia but also a city lot in what was described as the booming town of Huntington Beach, California, within a short distance of orange groves and overlooking the Pacific Ocean. Soon sets of the encyclopedia appeared on New England table tops from Bangor to Boston.

The California promoter sold out. So did the New England printer. Years passed. Grant deeds to Huntington Beach townlots were buried in the bottoms of old trunks; encyclopedias went to secondhand booksellers.

In June, 1920, Standard Oil found oil at Huntington Beach. The oil boom spread into the townlot area and lease brokers frantically searched out titles to townlots on which to drill wells. The search took them to New England and to the deeds that many had forgotten.

One New Englander, Ezra Hapfield, had purchased an encyclopedia for his daughter, Hattie, who was attending finishing school. When the deed arrived for a townlot in California, he put it in a desk drawer, later transferring it to an old trunk. Hattie married, bore a son, and, when her husband died, moved in with her father. Life on the farm went on.

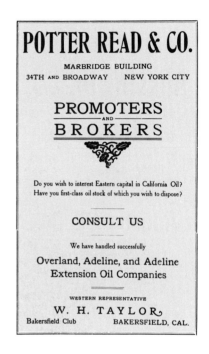

Eastern promoters and brokers turned to California for investment opportunities and focused national attention upon the California oil fields.

Black bonanza! Oil to make men's fortunes flows from a newly completed well at Coalinga, c.1902. (Standard Oil Company of California)

Unexpectedly a letter from California arrived. A firm of attorneys wanted information about the deed to a Huntington Beach townlot. Correspondence followed, and the firm offered more than three hundred dollars for the lot. The offer startled Ezra Hapfield, for it was more than he had paid for the encyclopedia and the lot. He paid up back taxes on the townlot, redeemed the property and headed for California. The lot, like others in the "encyclopedia section," proved to be over rich oil sand. With their first oil royalties, the Hapfields bought a house on a slope overlooking the sea, with orange trees in the backyard and winter-blooming roses on the porch. For them, as for others, royalties kept coming; unwanted books and unsalable beachlots made hundreds of thousands of dollars for unsuspecting buyers.

4 Life in the Valley Oil Towns

Oil booms had a great effect on town life in the San Joaquin Valley, transforming sleepy farming communities like Bakersfield into roaring centers of excitement and stimulating the growth of towns, such as Coalinga, in the midst of developing oil fields or, like Taft, at the end of railroad spur lines. Some of the valley's oil production pioneers, of course, came in ahead of the railroads, paved roads, frame houses, and adequate supplies of drinking water.

When Carl Baker and his wife of one year, the former Minnie Zumwalt of Willows, moved to Coalinga from Los Angeles in 1899, they had to leave their baggage twenty miles from the town. "There was no depot in Coalinga," Baker recounted, "so the old baggage man threw our baggage off at Huron—that was the last depot on the run—and then went on to Alcalde and turned the train around on the turn-table up there," three miles beyond the loading platform that served Coalinga.

The baggage man brought their luggage up the next day, and the Bakers went by wagon to the lease on which he was to drill a well for a man named Westlake Rummel for $2 a foot. "We got on a load of lumber, my wife and I, and went up Walker Canyon up around the hills, three miles west of Coalinga, Section 1, where the well was supposed to be drilled."

The barren hillside held no place to live, no derrick, nothing. Only the load of lumber and the two newcomers. "I built a little house about twelve feet wide and sixteen feet long," Baker recalled.

Mrs. Baker did the cooking. Carl Baker proceeded to build the derrick that would be used for the well. Living was primitive and without much comfort. Minnie Baker always said "that was the only time that she ever had a bedstead with only one leg," her husband remembered. How could a bedstead function with only one leg? "Built in the corner of the house, a bed only needs support on the odd corner," he explained.

Living quarters and derrick built, men hired, Baker began the job. "I took the contract to drill it to 1,000 feet, and if we

Taft, 1910; just beyond the Hotel Alvord can be seen a magazine stand, the O.K. Baths and Barber Shop, a steam laundry, and the C & C Theatre. (Kern County Museum)

didn't get any oil and got some gas, I'd drill it to 1,300 feet. I got the well drilled on Washington's Birthday, 1900," the first of many Carl Baker completed in the vicinity of Coalinga.

The settlement of Coalinga traced its beginnings, and its name, to early efforts at coal mining in the region. The first Spanish sheepherders and mustang runners in the area had warmed themselves by fires made of the low-grade coal they found in outcrops. At the end of the nineteenth century other men came expressly to develop a coal-mining industry. As matters developed, they had the right idea in seeking mineral wealth, but they sought the wrong mineral. They staked their money on developing coal mines and paid little attention to the heavy oil that seeped to the surface in outcrops, only collecting it occasionally for use in greasing the axles of their wagons. In the mines, the veins of rock that oozed oil were nuisances. Coal proved untenable as a mineral on which to base an industry, however, and the small community of Coalinga faced economic ruin in the early 1890's. Then the shift began from coal to oil. The completion of the Canfield well in 1897 started the oil development that Carl Baker helped advance.

The Coalinga field boomed and soon led the state in oil production. Along with big production, the field could lay claim to fame of another sort. The widest roads in the state led to the oil town.

"You'd start out over those deserts," Baker recalled, "and when the ruts got too deep, you'd just move over. The road from Coalinga was a half mile wide going out to the west side, and the same way going over to the east side. Early transportation was a chore because of the ten-mule and ten-horse teams and fourteen-horse teams. They just kept moving over, and the road kept getting wider because the ruts would get so deep you couldn't drive in them."

The oil boom had changed the empty appearance of Coalinga by February, 1908, when Denny Driscoll, a Pennsylvania machinist, arrived, but the town remained wide-open. Driscoll stepped from the train at the new Coalinga depot right out onto "Whiskey Row," officially known as Front Street. The day was untypically wet, and the sidewalk stood only three

Front Street, Coalinga, 1912; better known as "Whiskey Row." (R. C. Baker Memorial Museum, Inc.)

feet above the ooze that constituted the street. Gangs of oil men playfully tried to push each other over the edge of the boardwalk into the morass. Every time one man took a dive, a roar of laughter went up from the crowd that solidly packed the block. Driscoll dodged the crowds, telephoned his new headquarters at California Oilfields Ltd., and a buggy presently arrived to drive him nine miles out to camp where a cottage awaited him.

Later he ventured back to Whiskey Row, where establishments like the Portola Bar and Grill, the Axtell, and the Palace served Tom and Jerries, beef tea, clam juice, hot scotch, and oyster cocktails to their clientele. "It was a solid block of saloons in a wide-open town that never closed down," Driscoll recalled. "Every establishment in the block had several tables of poker and blackjack constantly in progress. The dance halls were across the tracks directly opposite the Row, and they, too, ran wildly and uninterruptedly. The City Marshal mingled with the crowds but had little to do as

no one ever stepped out of line. The men were boisterous, but law abiding.

"It amazed me at first," Driscoll said, "to see the stacks of twenty, ten and five dollar gold pieces piled up like poker chips in front of the players, and it wasn't uncommon to see a whole stack of these gold pieces shoved to the middle of the table on a single bet. We rarely saw a gold piece in the East, and this affluence fairly made my eyes pop. When I left Pennsylvania we were being paid our wages in scrip and any kind of money was as scarce as hen's teeth. Strangely enough, there was no lawlessness in Coalinga at this time. Women were quite safe on the streets at any time of the day or night—even the dance hall girls.

Oil men in Coalinga meet to socialize in a town bar and grill. (R. C. Baker Memorial Museum, Inc.)

Wagons and automobiles take men to work at California Oilfields, Ltd., Coalinga. Photograph by Frank Foster. (R. C. Baker Memorial Museum, Inc.)

"Coalinga was a strange town in those days," Driscoll concluded. "It had the atmosphere of an impending storm, but everything remained peaceful. There was a little noise but nothing of a riotous nature took place. It boasted some of the finest people in the world. Men trusted each other. Nobody ever locked a door. You could make a deal with anyone, stranger or otherwise, secure in the knowledge that the terms would be rigidly adhered to."

The town had a growing commercial community, including several leading oil supply firms. It also had a women's club, founded in 1905, whose members undertook the establishment of a free reading room "especially for men" that antedated the town's public library. The group also became influential in the movement that secured a high school for the growing town. The science editor of the *Overland Monthly*, one of California's leading magazines, visited Coalinga in 1908 and had nothing but praise for the community. "The stores of Coalinga are metropolitan in appearance and well stocked," he reported, "and there is a prevailing air of independence and patriotism. The great Coalinga oil field is being developed by Californians and in regular California style."

Enterprise and civic pride also characterized the burgeoning town of Taft, incorporated in 1910. The community first took shape in 1908 when the Southern Pacific and Santa Fe railroads extended their jointly owned spur line in the West Side

oil fields to take in developments at Midway. They named the railroad siding seven miles northwest of Maricopa, "Siding Two," and around it clustered a lumber yard, supply houses, and saloons. Another cluster of tents and saloons became known as Moron; both settlements coalesced into the new community of Taft, named in honor of the incumbent President of the United States. The name, "Moron," intended no political commentary; some folks had hoped to call the new settlement "Morro," but when it became known that California already had a town named "Morro Bay," the name became subtly altered.

In 1910, a group of women undertook to name the streets in Taft, and several business-minded citizens opened new tent hotels on Center Street. A night's lodging cost thirty-five cents, and a good meal could be purchased for a quarter. Living facilities remained primitive, however, in the dry desert country of the developing Midway-Sunset field. Few buildings existed, and field crews lived in tents or box houses with cracks in them "big enough to throw a dog through," recalled Lindsay (Lin) Little, a South Carolinian who came west with a party of southeastern drilling experts in 1908. The region had scant vegetation, and its prime inhabitants seemed to be rattlesnakes, centipedes, and tarantulas, the newcomers thought. Drinking water proved to be a much scarcer commodity than oil and, at one dollar per barrel, cost twice as much. Paying five cents a glass for drinking water became standard practice in Taft.

Newcomers like Little may have profited from the advice offered by the new weekly magazine, *California Oil World*, which Charles P. Fox had begun to publish in Bakersfield. Its first issue in 1908 contained a column of hot weather advice. A "wise physician," it seems, advised that "wherever possible work be done in the shade, and that the clothing be worn as loosely as circumstances will permit. A diet restricted almost entirely to fruits and vegetables with very little meat will be found most capable of being handled to advantage by the system." As for liquid intake, the column prescribed an expensive medicine—water.

The community had a frontier aura, emphasized by the

Siding Two on the spur line serving the West Side oil fields, 1909. This cluster of supply houses, lumber yards, and saloons incorporated into the city of Taft in 1910. (Robert B. Moran)

Fourth of July celebration, Taft, 1911. (William Rintoul)

"Old Betsy," a steam-driven Best tractor with a train of trackless cars, made two trips a week between California Oilfields Ltd. camp and Coalinga to pick up heavy oil field equipment. (R. C. Baker Memorial Museum, Inc.)

region's inhabitants in the cars they drove, even as cowboys in other areas had once placed great reliance on their horses. Clarence Berry, a prominent oil operator, made the news when he arrived in the West Side fields driving a seventy-horsepower American roadster, promptly named the "Red Devil." The car had forty-inch wheels and was said to be capable of doing eighty miles per hour. It was also news four days later when Berry was arrested twice in Bakersfield, charged with speeding from the Southern Hotel to his garage.

Even after World War I, when Taft and the other valley oil towns had become well established, primitive threats still lurked. The famous mouse invasion of 1926–27 became especially memorable, as recalled by Ronald W. Heath, who spent his Stanford summer vacations working in the valley oil fields before going to work full-time with Signal Oil and Gas. The mouse invasion originated in the abundant grain crops local farmers had planted in the fertile land of dry Buena Vista Lake. The rodents fed on the grain and multiplied quickly; when the rains returned to the area, the rising water drove the mice in all directions. They swarmed into Taft "by the millions, invading homes, stores, warehouses, doghouses, and even gasoline stations in their frantic search for food," recounts Richard Sneddon, noted oil industry chronicler who interviewed Heath and many other oil production pioneers.

"Oil operators," reports Sneddon, "dug trenches around the drilling rigs and office buildings and poured poisoned grain into them, but it was a losing battle." In a three-day

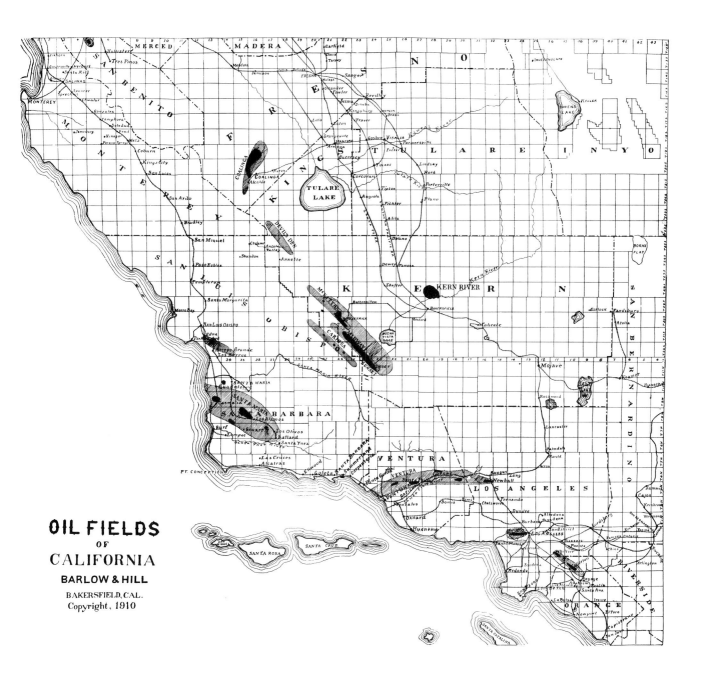

OIL FIELDS
OF
CALIFORNIA

BARLOW & HILL

BAKERSFIELD, CAL.
Copyright, 1910

71

period one trench alone accounted for more than 75,000 mice, among them, according to an account in *The Bakersfield Californian,* "genuine rats wearing shaggy winter coats." The leases that had horses and stables with feed became special targets, and the local highways became carpeted with mice pelts. The stench grew insufferable, and it took some time before a government extermination expert named Piper— Stanley E. Piper—arrived to take over the campaign against the "plundering horde."

Piper set up a base camp on Pelican Island in the northern portion of the dry lake, outfitted the camp with living quarters and cookhouse, and recruited a force of twenty-five men— promptly dubbed the Mouse Marines—to carry on the battle with the rodents. More than one thousand ring-billed gulls appeared to aid the exterminators, diving at the mice along with companies of ravens, hawks, great blue herons, and other airborne assistance. In mid-February, 1927, the great mouse war ended with losses calculated at more than thirty million mice and at cost of $10–15,000 for damaged crops and poisoned grain.

Bakersfield had become the valley oilmen's metropolis, the

Married employees could secure comfortable company houses at low rentals; the happy couple in this Frank Foster photograph lived at California Oilfields, Ltd. Coalinga camp. (R. C. Baker Memorial Museum Inc.)

center of speculation and reports of new developments, not only in the San Joaquin Valley but around the world. Men would leave Bakersfield on assignment for their companies in South America or the Far East and return with news of oil prospects in these regions and the cosmopolitan aura that foreign travel provides.

The town aspired to eminence in cultural and sporting affairs as well, boasting vaudeville shows, circuses, boxing matches, and other entertainment not provided in the smaller valley towns. A touring company of the Ziegfeld Follies of 1910 played the Bakersfield Opera House, and for a ticket price ranging from fifty cents to two dollars local theater-goers could see the seventy-five Anna Held girls and a "formidable cast of stars," including headliner Bert Williams and a young singer named Fanny Brice. Some years earlier the opera house stage had been the scene of an exhibition boxing match featuring Jim and Jack Jeffries, and on another occasion the world's wrestling champion, Frank Gotch, made an appearance there.

Amateur athletics also had a great deal of attraction. One famous event in the early boom days was the footrace from the Kern River field to the Southern Hotel, pitting an oil worker known as the Road Runner from the Four Company against the Kentucky Kid from the West Shore Company. The Road Runner outdistanced the Kid by fifty yards, covering the seven miles in forty-six minutes. Reportedly, more than $700 in bets changed hands.

Though the discovery of oil at Kern River had changed the town from a sleepy farming community to an active oil center, many residents felt Bakersfield had not received the national recognition that was its due.

To remedy the situation, a group of Kern River oil men put forward a scheme in the spring of 1911 which they said would make the city known from coast to coast. They proposed that a 150-mile automobile race be held over a course to be constructed through the Kern River oil field and the outskirts of the city, with entries solicited from the fastest cars in the country. The time to hold the race, they suggested, would be on the Fourth of July, normally observed with riotous vigor in the oil fields.

Local reaction to the proposed race was enthusiastic. A. Weill, a prominent businessman, threw merchants' support behind the race, stating that it "would advance the city and also would show country customers that local merchants wished to give them entertainment on a holiday in recognition of their patronage." In the oil fields, there was keen interest in how fast a thing could be done, whether it was drilling a well or driving an automobile, and the race captured the imagination.

There were, of course, problems to be overcome. Especially the matter of prize money. Planners decided the winner would receive $2,000, second place $1,000, and third place $500. It was one thing to agree on amounts, however, and another to raise the money. At a meeting, Angus Crites, Jim Bruce and other oil men from the Kern River field pledged $1,000; others present pledged a total of $720. *The Californian* hailed the Kern River oil men as "live wires" and called on others to join the fund-raising campaign with an editorial that stated:"Everyone who has civic pride, love of the home place and who wants Bakersfield to take high ranking as a city of enterprise in a county of opportunity, will appreciate a chance to lend aid to the great undertaking." The local newspaper also predicted that the race would bring 20,000 visitors to town, a number substantially higher than the 12,727 residents credited to Bakersfield in the 1910 census.

The proposed race course existed in large part only on its promoters' maps, and it took a great mobilization of effort to secure rights of way, arrange with the Southern Pacific Railroad to halt traffic on the spur line crossing the race course, build bridges, and straighten and realign sections of roadway. The Kern River oil men went to work with characteristic enthusiasm, and they managed to secure sanction for the race from the Automobile Association of America.

Various oil companies put more than 200 men to work on the course, and dust rose in clouds along the route. Workmen commandeered virtually every watering wagon in the Kern River field to sprinkle water to keep down dust. When more water was needed at one distant point, pipeliners laid more than a mile of new line in less than a day.

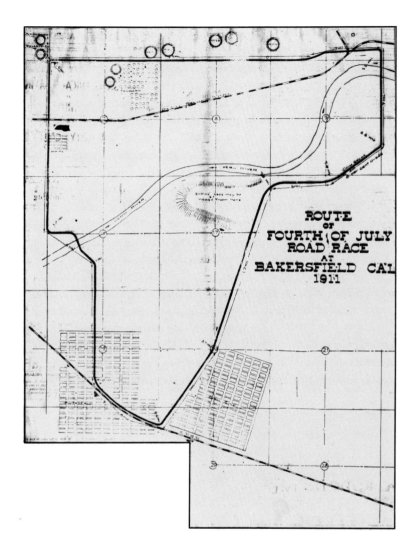

ROUTE
OF
FOURTH OF JULY
ROAD RACE
AT
BAKERSFIELD CAL
1911

The race course ran from town out to the Kern River oil field. Spectators were advised that the entire race could be viewed from the 300-foot high river bluffs, a spot marked in the center of this map. (Beale Memorial Library, Bakersfield)

Kern County merchants sold promotional badges for twenty-five cents each in order to raise funds, and local theaters held benefit shows. Carloads of "Bakersfield Boosters" took to valley roads to tout the race. A box on the newspaper's front page counted the days until the race, and word came into town of various famous drivers and automobiles that might participate: Harvey Herrick, Bert Dingley, and Frank Seifert.

Harvey Herrick made a trial run around the course in a National 40 and said, "The track is about the roughest I ever saw." Dingley arrived with a fifty-horsepower Pope-Hartford racer which he claimed could reach speeds of 100 miles per hour on good stretches. *The Californian* reported that the men of the hour in Bakersfield were "the men with tight leggings, goggles, and helmets."

Parts of the race course served as general thoroughfares in town, but they were reserved for exclusive use of the racing cars from four to six each morning. On the day of the race, work came to a standstill in the Kern River oil field, train traffic stopped on the spur line, and a crowd estimated at 10,000 persons lined the bluffs and the path of the race. Dingley's Pope-Hartford racer was the bettors' favorite.

At 11:05 a.m., the starter waved off Frank Seifert behind the wheel of a Mercer, followed by Harvey Herrick in the National, Louis Nikrent in a Marquette Buick, and Bert Dingley in the Pope-Hartford. The blue National took the first lap in 11:58.2. By the end of the third lap, the Buick had blown a plug and was in the pit for seventy-four seconds; the car went out at the end of the fourth lap with a broken connecting rod. Seifert

Harvey Herrick, driving this National 40, wins the Fourth of July race, 1911. (Kern County Museum)

in the Mercer, outpaced by the National and Pope-Hartford, realized that with the Buick out of the race he had only to finish to win third prize of $500; he settled down to survive without attempting any speed records. The race became a two-car trial between the National and Pope-Hartford.

Dingley in the Pope-Hartford began coming on in the fourth lap, taking it in 12:09.8. The track was dusty, and the going became progressively worse. The National blew a tire in the sixth lap; the Pope-Hartford gained two minutes. The Pope blew a rear tire in the eighth lap and went into the pit for a tire change. Herrick in the National roared ahead, making the tenth lap in 11:18.2, the best time cf the day. Keeping the car wide open, Herrick roared on to hold the lead the rest of the way, finishing the fourteenth and final lap in a total time of 2:58:58.5. Judges declared the race ended, awarding second prize money to the Pope-Hartford, which had finished eleven laps, and third to the Mercer, which finished seven.

At a banquet at the Southern Hotel, victors and vanquished drank champagne from a $1,000 silver trophy cup donated by the Tevis brothers. The city attorney predicted that Bakersfield would become the racing center of the west; *The*

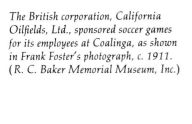

The British corporation, California Oilfields, Ltd., sponsored soccer games for its employees at Coalinga, as shown in Frank Foster's photograph, c. 1911. (R. C. Baker Memorial Museum, Inc.)

Children of oil workers went to school in the midst of the oil fields, as at Sunset in 1902. Gordon P. Suiter, donor of this photograph, is pictured in the first row at right. (Petroleum Production Pioneers Collection, Long Beach Public Library)

Californian described the Fourth of July, 1911, as "Bakersfield's greatest day," and Harvey Herrick announced plans to challenge the world's mile stock car record.

Civic pride and excitement, along with the spirit of hustle and a desire to get things done without delay, characterized all the San Joaquin oil towns. Underlying the situation, however, was one unwritten rule: not everyone could participate. While the oil fields were a friendly place for most white people, racial discrimination prevailed there as it did throughout much of the country. Most people believed that the color of a person's skin made a difference, and whites believed that black, brown, and yellow peoples were inferior. Such racist attitudes enabled a Bakersfield newspaper to report the sinking of a Canton steamer under the headline, "300 Chinks Are Lost at Sea," and to describe the lynching of a black youth in Kentucky under the heading, "Another Negro Fiend Lynched."

When a road show company came to Bakersfield to present a performance of "The Clansman," with a cast of seventy-five, a committee of concerned black people appeared before the town's council to protest Thomas Dixon's Southern play, stating that it would only call up race hatred. The show went on as scheduled, playing to a capacity house, but the newspaper's reviewer found it "a useless and unnecessary portrayal." On the next page of the same issue of the paper a dispatch from Union City, Tennessee, reported the lynching of three Negroes.

In oil towns like Taft and Coalinga and others, racial feelings were exaggerated to the point that these places became widely known as "white man's towns" in which blacks were not permitted to remain overnight. Nor were there many Orientals, and those who lived in the oil towns normally were confined to roles as gardeners or cookhouse help. Denny Driscoll recalled that when he arrived in Coalinga "colored porters were not permitted to travel beyond the confines of the station platform, and they stayed on the trains overnight."

In Taft, the situation took an ironic turn in the fall of 1913

Derrick Avenue, near Maricopa in the Sunset portion of the Midway-Sunset field, 1912. (J. J. Oliphant, Petroleum Production Pioneers Collection, Kern County Museum.)

A Maricopa wedding party posed for its portrait in the town's main street in 1909. The photographer managed to line folks up in a way that gave prominence to his sign (upper left). (Standard Oil Company of California)

when the town, in an effort to attract national attention, became the setting for a boxing match that would name a challenger to the black heavyweight champion, Jack Johnson. The match pitted Jack Lester, an oil field favorite, against one of the country's leading heavyweights, a black boxer named Sam Langford. Though blacks were not permitted to remain overnight in Taft, an exception obviously had to be made in Langford's case, and with his manager and sparring partners, he was put up in a cottage on Center Street, doing his training in the Mariposa pool hall and his roadwork through the oil fields, where sporting interest temporarily at least overcame prejudice and he was enthusiastically greeted by oilworkers.

As interest mounted in the boxing contest, Norbig Film Company of Los Angeles announced plans to film the blow-by-blow action for worldwide showing, and an excursion train was arranged to bring in fight fans on the day of the match. This posed a problem, for among the fans were a group of black people in a car of their own. Since the train would arrive at lunchtime, where could the blacks eat? Promoter George Wilson, mindful of Taft's reputation as a white man's town,

Southern Hotel, Bakersfield; headquarters for visiting oil men, whose comings and goings received full coverage in the local paper. (California Historical Society Library, photograph.)

took steps to solve the problem. As blacks left the train on the day of the fight, they were advised that if they cared to eat while in Taft, they would be welcome to do so at the local Chinese restaurant.

The fight itself was overshadowed by the collapse of a portion of the newly-completed bleachers which plunged more than 100 people twenty feet to the ground. Two men were injured seriously enough to require hospitalization. The boxing match proved to be a mismatch, and the talented Langford remained the leading contender by beating Lester in four rounds. As soon as the fight ended, Langford and his entourage—their bags packed before the contest—jumped into a seven-passenger Studebaker owned by Al Israel, proprietor of the Crawford Bar, and sped off to catch the train out of Bakersfield.

5 Oil in the Streets of Los Angeles

In contrast to the vast expanse of the valley oil fields, wooden derricks crowded residences in Los Angeles for space, and the city had something of a forest without ever having had many trees. As a matter of pride one major area of drilling within Los Angeles itself became known as the City field, and it lay across the area now occupied by Dodger Stadium and portions of the Hollywood and Santa Ana freeways.

Edward L. Doheny, a mining prospector down on his luck, completed the Los Angeles discovery well for seven barrels a day in November, 1892. Doheny had observed the Angelenos gathering *brea* or asphaltum from the area's tarpits for use as fuel in coal-scarce California. Realizing that this crude tar was petroleum that had congealed upon contact with the open air, as pine sap turns to resin, Doheny explored the residential neighborhood near Westlake Park, pooled resources with Charles A. Canfield, an old mining crony, and purchased a city lot for $400. Unaware of oil drilling methods, they began by sinking a four-by-six-foot miner's shaft, digging it out by hand with picks and shovels. They found an oil seep seven feet below the surface and kept digging despite the presence of gas. They finally gave up at 155 feet, nearly overcome by the fumes at that depth. Doheny then fashioned a crude drill from a sixty-foot eucalyptus tree trunk and continued to bore the hole. On the fortieth day of work, gas burst out of the hole, and oil bubbled up into the shaft. Doheny proclaimed a new day in the city's economic life, and the boom was on.

With fortunes to be made, the residential district became crowded with promoters, drillers, and derricks. Trampled gardens, chugging and wheezing pumps, flooded lawns, and other nuisances went along with the attempt to turn backyards into paydirt. In an area bounded by Figueroa, First, Union, and Temple streets, more than 500 wells were producing oil by 1897, and wildcatters exploring for new drilling sites moved slightly northwest into what would become another part of the City field.

Some of the first men attracted to oil development in the Los

Wooden derricks crowded residences for space in Los Angeles, Court Street, 1901. (Huntington Library)

Angeles City field were veterans of the Pennsylvania rush, among them Marcellus Manley, a self-educated orphan who had taken time out from attendance at Ohio Wesleyan College to see duty in the Civil War. Following his discharge from the Union Army, he had worked in the oil fields in Pennsylvania before returning to college to finish his education. After graduation, he taught at his alma mater. While still teaching there, he accepted a position as principal of the high school at Santa Ana, California, and brought his family west. While serving as principal, he started the first public library in Santa Ana.

In 1896 he traveled from Santa Ana to Los Angeles with the idea of getting into the oil business by buying up existing oil

Before 1901 there were more than 1,000 wells in the northwest part of Los Angeles. Before the construction of pipelines, the horse-drawn oil wagon was a familiar sight. (Historical Collections, Security Pacific National Bank)

(overleaf): State Mining Bureau map of the City field, Los Angeles, 1903. The Manley property at Sunset and Beaudry is at the circle, top center.

companies. Buying companies was a relatively simple affair. According to Manley's grandson, Kenneth Manley, "When a person owned or leased a lot in the Los Angeles oil field area, he merely hunted up somebody who was supposed to be a driller or had some equipment and drilled a hole. Of course, there were hundreds of one-and two-well companies at that time," Manley said, and his grandfather consequently had no trouble buying up wells and companies to start Manley Oil Company, a firm that would last through three generations.

The first oil produced in the City field was used in efforts to pave the local streets. The oil settled the dust, but, as Kenneth Manley explained, "this ended up being a gummy mess which stuck on the carriage wheels and on everybody's feet and was not practical at all." In 1902 a group of oil operators persuaded the City Council to authorize the experimental use of an oil burner in the building now occupied by the May Company, and with the success of that project the market for oil as fuel became established. Additional impetus came shortly thereafter when railroads converted wood- and coal-burning locomotives to oil, greatly expanding the market for petroleum products. Companies proliferated in the Los Angeles City field. Shortly after the turn of the century more than 140 companies produced oil in the field; a decade earlier only four companies had been producing oil in the entire state. Many of the field's producers had only one or two wells, but Rex Crude Oil had sixty-three and Parker and Los Angeles Transfer and Terminal companies each had more than forty.

Drilling wells in the Los Angeles City field posed the problem of making oil production compatible with urban living. Noise, dirt, traffic, odors, and waste disposal all had to be dealt with, and at least one of the solutions proved unique. One man had a rig in his backyard, and no place for a sump in which to run the waste water and mud. However, his house had a basement. That's where the mud went.

Manley Oil Company had six wells on a property near Sunset and Beaudry where St. Vincent's Hospital used to stand, on a spot later to be occupied by the Metropolitan Water District. The company had a small drilling rig with a fifteen-horsepower White & Middleton gas engine that powered

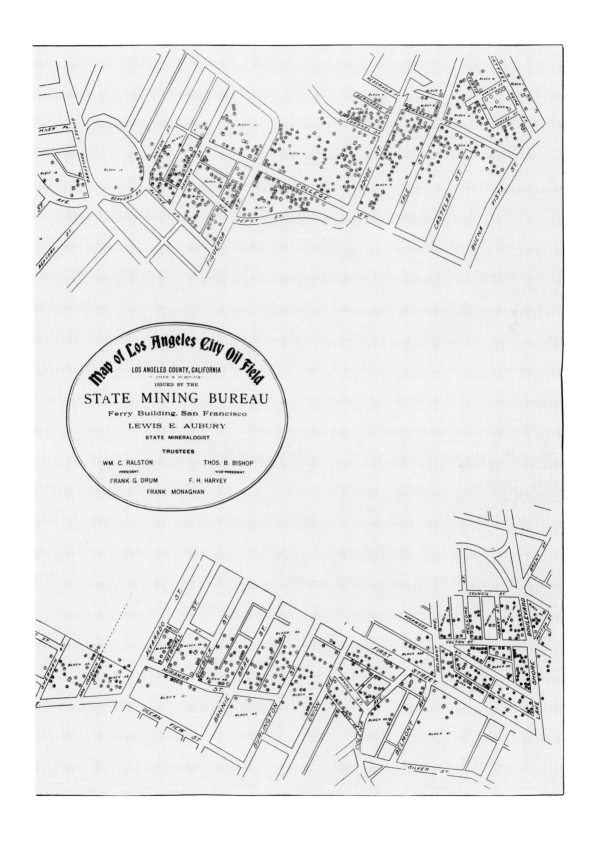

Map of Los Angeles City Oil Field

LOS ANGELES COUNTY, CALIFORNIA
By CHAS. A. BLACKMAN

ISSUED BY THE

STATE MINING BUREAU

Ferry Building, San Francisco

LEWIS E. AUBURY

STATE MINERALOGIST.

the rig. One of its flywheels came loose, but the men working on the rig did not notice until the engine began to make a strange noise. They looked back just in time to see the flywheel spinning off down Sunset. The flywheel rolled to where the Pasadena Freeway now runs underneath Sunset Boulevard. There it fell over and stopped.

Later, Kenneth Manley, hearing the story recounted by his father, asked what they did. The elder Manley replied, "There was nothing to it. We took a sled down there and pry bars and put the flywheel back on the sled and towed it back up the side of the hill and put it back on the engine." Recalling the story many years later, Kenneth Manley added, "I shudder to think of what would happen now, half a century later, if one of those six-foot flywheels went sailing down Sunset Boulevard during rush hour traffic."

Horses created additional traffic hazards in the City field neighborhood, Manley recalled. A new horse had been broken into a four-horse team pulling eighteen-barrel wagons.

E. L. Doheny brought in the Los Angeles discovery well in 1892. In the 1920's the event was reenacted with appropriate ceremony, and Doheny stood with his arm held high as he proclaimed the new day in the city's economic life. (Huntington Library)

Oil fields near West Lake Park, Los Angeles, 1906. Photograph by Ralph Arnold. (Huntington Library)

One morning a teamster was engaged in loading oil from a 600-barrel wooden tank onto the wagon. "Right about then," said Manley, "the fire department gang went clanging by. That was all the new horse needed. He started to rear and whinny and plunge, and finally, he took off with the wagon and the rest of the team." This yanked the lead line out of the tank, leaving a large hole. Keeping his presence of mind, the teamster pulled off his coveralls and poked them down into the tank opening, more or less slowing the leak. Meanwhile, Manley concluded, the team "charged through what is now Chinatown. As they made a left turn at College and North Broadway, the wagon overturned and the oil and everything else went flying into Little Joe's Restaurant."

The incident with the ex-fire horse emphasized the shortcomings in the early handling of crude oil. In fact, so much often was spilled around well sites that the oil on the ground had a tendency to soften horses' hoofs, making it not always feasible to use them. Some enterprising folks profited from the leaks, however. One of the city's manufacturing plants,

located on North Broadway about a quarter mile north of College Street, had a sump out in back that proved equal to one of the best oil wells in the Los Angeles City field. The field operators' tanks leaked, and rain washed the leaked oil down behind the plant. There the plant operators picked the oil up with a pump and sold it back to the field operator.

One of the hazards of trucking oil along Lakeshore Boulevard, which later became Glendale Boulevard, was the barrier of Echo Park Lake, the only natural lake in Los Angeles. In wet weather, the lake used to run all the way down to Fifth and

Oil collecting in vacant lots along Los Angeles streets, 1895. (History Division, Los Angeles County Museum of Natural History)

Flower streets. In the winter, the lake became such a sticky mess that it was sometimes necessary to hook up six horses to team the oil wagons around the lake. In 1907 operators got so much oil on the lake that it caught fire, burning for several days. Fortunately, the city did not have a smog problem then, and people were not too concerned. The oil burned off the lake, and teamsters went back to business as usual.

More than a thousand wells were drilled in the Los Angeles City field, causing the price of oil to drop at one time to as little as ten cents a barrel. One day a producer with wells and a storage tank on the side of a hill northeast of the intersection of Glendale and Beverly boulevards was busy near the tanks when a man came up and looked at the tanks. When asked what he wanted, the man said, "I'm from the City of Glendale, and we would like to buy a wooden storage tank, like this one."

The oil operator asked how much they would give for such a tank, and the man quoted a price.

The oil man's ears pricked up, and he quickly offered, "What about this one right here?"

The man said, "Well, let me climb up and look at it."

He climbed the ladder and examined the tank, then said, "This is about what we wanted and it would be just fine, except it's half full of oil."

"That's all right," the oil man said. "When do you want to come after the tank?"

The man said they could probably have the crew there the next day. "How soon would the tank be empty?"

"I'll empty it right now," the oil man said, and that's exactly what he did. He opened the valve and let the oil run down the street at the intersection of Glendale and Beverly.

Much ingenuity went into ways of getting oil out of the ground, but equally taxing endeavors became necessary to move the oil to refining and marketing centers. In the first decade of the twentieth century several major pipeline projects caught men's imaginations; Standard built a 300-mile line in 1902–3 along the Santa Fe Railroad's right-of-way from the Kern County fields to a refinery at Point Richmond on San Francisco Bay. Associated Oil and the Southern Pacific in 1908 completed a pipeline running between Bakersfield and the rail terminus at Port Costa on the Carquinez Straits. In 1909 the 280-mile Producers pipeline system, undertaken in association with Union Oil, linked the San Joaquin Valley with the coast at Port Harford on Morro Bay.

Three million dollars was spent on Standard's line. Eight-inch pipe was unloaded from railroad cars at yards containing twenty-five or thirty miles of pipe. Ordinarily gangs strung pipe at the rate of a mile or two a day, but on one memorable occasion, two gangs strung four and a half miles in a single day. Labor turnover was high, and a construction boss is said to have observed that he had three gangs—one coming to work, one working, and one coming in to be paid off. Pumping stations to heat and boost the oil through the line were built about thirty miles apart, and a great deal of experimentation proved necessary before effective means of heating the heavy crude oil were established to permit an easy flow through the line.

The pipeline built by Associated Oil and the Southern Pacific employed a new type of pipe which was supposed to move heavy crude oil without the necessity of heating it first. The so-called "rifled" pipe, invented by Buckner Speed of Berkeley, was designed to reduce friction when pumping heavy crude oil. Line pipe was corrugated and then given a twist of one complete revolution in every four feet, thus producing a rifling effect.

By means of separate pumps and special injection nozzles, oil and water were forced into the line at the same time. The

Producers pipeline crossing the rugged coastal hills en route to Port Harford near San Luis Obispo, 1909. (Union Oil Company of California)

water worked as a lubricating agent by forming a jacketing layer between the oil and the pipe. The rifling was expected to maintain the two in the same relative positions throughout the run. In theory, the viscous low gravity crude oil would not touch the inner walls of the pipe and contact would be by water only, thus markedly reducing line friction. Because of the lowered friction, pumping stations could be spaced much farther apart, and so eventually a great saving in transportation cost might be achieved. The system worked perfectly on short runs, but on long runs under high pressures, many difficulties arose, not the least of which was the tendency of crude oil and water to mix into an emulsion that was almost impossible to process. The basic premise of the rifled line —the introduction of water to serve as a shield between the oil and the inner wall of the pipe—had to be abandoned. The Associated pipeline had to resort to the hot oil principle and increase the number of pumping stations. In 1910, a second line, equipped to pump hot oil, was completed between Maricopa and Port Costa, paralleling the first line from Mendota north.

Standard Oil refinery under construction near Richmond, California, 1901. (Standard Oil Company of California)

At right: Port Costa, terminus of the Associated pipeline. (above) 1914 view shows Burlington Hotel, Power house, and railroad; (below) Balfour Guthrie California Warehouse and Dock Co., 1905. (Louis L. Stein, Jr.)

The concept of the Producers pipeline originated in a complex series of marketing maneuvers that pitted the numerous independent producers in the San Joaquin Valley against the large Standard Oil and Associated Oil companies.

By 1909, California had become the leading oil-producing state in the nation. The bulk of the state's production came from the San Joaquin Valley, where 150 small companies, unhappy with the prices offered for their crude oil, banded together as the Independent Oil Producers Agency. Among them the companies produced a large share of the fifty-two million barrels of oil sent to market that year from California. Neither Standard Oil nor Associated Oil, the two largest bulk buyers, would pay the producers what they thought they should receive, so L. P. St. Clair, spokesman for the independents, long on friendly terms with executives of Union Oil, asked the Union management to take over as sales agent for the organized smaller independents.

A ten-year agreement resulted, whereby Union undertook to handle all the output of the Producers Agency, guaranteeing it the same price that Union got for its own products. Nearly all of the oil represented in the agreement came from San Joaquin Valley fields, where the high cost of rail transportation had prevented Union from matching its two competitors, Standard and Associated, both of whom had pipelines.

To move the oil to the sea, Union and the Producers Agency jointly organized the Producers Transportation Company to

Crews of strong men work on laying the Producers pipeline in 1909. (Union Oil Company of California)

Oil fields and oil pipelines of California, 1921; at right.

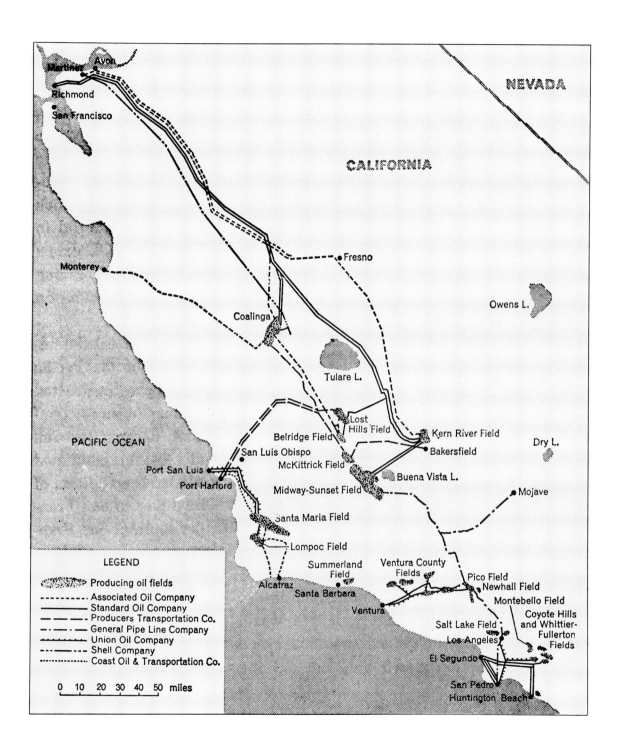

NEVADA

CALIFORNIA

Martinez · Avon

Richmond

San Francisco

Monterey

PACIFIC OCEAN

Fresno

Owens L.

Coalinga

Tulare L.

Lost
Hills Field

Belridge Field

Kern River Field

Dry L.

San Luis Obispo

McKittrick Field

Bakersfield

Port San Luis

Buena Vista L.

Port Harford

Midway-Sunset Field

Mojave

Santa Maria Field

Lompoc Field

Summerland
Field

Ventura County
Fields

Pico Field

Newhall Field

Alcatraz

Santa Barbara

Montebello Field

Coyote Hills
and Whittier-
Fullerton
Fields

Ventura

Salt Lake Field

Los Angeles

El Segundo

San Pedro

Huntington Beach

LEGEND

Producing oil fields
- - - - - - Associated Oil Company
————— Standard Oil Company
— — — Producers Transportation Co.
—·—·— General Pipe Line Company
——+—— Union Oil Company
—··—··— Shell Company
·········· Coast Oil & Transportation Co.

0 10 20 30 40 50 miles

move oil from the San Joaquin Valley to the coast at Port Harford, now known as Avila, just south of San Luis Obispo. The pipeline the company decided upon was a tremendous undertaking for the time; it called for 280 miles of pipe, fifteen pumping stations, field tankage to store twenty-seven million barrels of oil, and wharf facilities at Port Harford. Original estimates figured the cost of construction at $3 million; by the time the line was completed, however, Union Oil and the Agency members had spent $4.5 million.

The Producers group designed the line to serve a number of fields in the San Joaquin Valley, notably Kern River at Bakersfield; Midway-Sunset, which embraced the communities of Taft, Maricopa and Fellows; McKittrick, by the community of the same name; and the Coalinga field some ninety miles to the north. Oil would be delivered to a tanker-loading facility at Port Harford and then shipped to San Pedro or Hueneme for ten cents a barrel, San Diego for twelve and a half cents a barrel, or north to San Francisco for ten cents a barrel and Eureka for twenty cents a barrel. A charge of twelve and a half cents a barrel was to be made for transportation

Teams of horses hauled boilers across the desert for pumping stations along the Producers pipeline, 1909. (Union Oil Company of California)

Workmen raised the boiler smokestack through the boilerhouse frame during construction of the Mack station on the Associated pipeline, 1908. (Getty Oil Company)

through the pipeline; there were to be storage charges of two cents per year per barrel in earthen reservoirs or one cent a barrel per month for storage otherwise.

The system consisted of eight-inch pipe except for a short section from the Sunset district at Maricopa to the Midway district at Fellows, which was six-inch line. The pipeline traveled from the Kern River field west to the McKittrick pump station with intermediate stations at Rio Bravo and Buttonwillow; two lines from Sunset to McKittrick via Midway station; and two eight-inch lines from McKittrick north to Junction via Middlewater Station. At Junction an eight-inch line joined the stream, carrying oil from the Coalinga fields via Tar Canyon and Dudley stations. Thence the line snaked west to Avila, via Antelope, Shandon, Creston, Santa Margarita stations and San Luis Obispo tank farm.

Where two eight-inch lines ran, the plan directed that one line would handle heavy crude and the parallel one would take care of the light crude. All stations employed steam to power the pumps, with exhaust steam being used to heat the heavy crude and reduce its viscosity.

This Heine boiler was shipped by rail to Coalinga, where it was unloaded for hauling to Associated's Mack pipeline station. (Getty Oil Company)

Four gangs labored to build the line, one starting from Coalinga, a second from the coast end, another from McKittrick, and a fourth from Junction. Each gang comprised 100 to 125 men, with an almost equal number in the crews digging the trenches. Eight- and ten-horse teams lined the route; Producers Transportation Company employed 540 head of horses for the construction effort. The company paid $14 a day for an eight-horse team and wagon, fed the driver, but required that he furnish feed for his stock.

Fritz Karge, who directed pipeline operations for Union Oil from 1915 to 1949, explained that the pipeline project was carried out "under the direction of a Texas pipeline contractor who had been accustomed to laying lines for the transportation of light Texas crude oil which flowed more easily than the heavy California crude oil. The pump stations were consequently designed, equipped and spaced with the easier flowing commodity in mind. Some of the stations had one or two large high efficiency flywheel pumps, while others had straight line units which, although less efficient, were less costly. Distances between stations were relatively uneven although the profile of the line route in many areas was quite uniform," he said.

The winter of 1909 in the San Joaquin Valley was an exceptionally wet one, Karge related. "At the proposed location of the Rio Bravo and Buttonwillow stations veritable lakes formed—and were quite likely to form again should the same weather conditions prevail." The site of the Rio Bravo station consequently had to be moved three miles to the west, lengthening the route from Kern River; and the Buttonwillow site was moved two miles to the east, increasing the distance to McKittrick. "Here also at one spot the profile of the route rose to an elevation of 600 feet, adding decidedly to the load at the Buttonwillow station," Karge explained.

"As a result," he continued, "when pumps at Kern and Buttonwillow were held at maximum allowable pressure, the volumetric flow was considerably reduced and was even worse under adverse weather conditions. Attempts were made to solve the problem by digging trenches under the line in the coldest areas, pouring gasoline or other fuel therein, and igniting this to heat the line. However, this was costly, dangerous and ineffective," Karge emphasized.

Hardly had the pipe-laying job started before the company ran into trouble with the Board of Trustees of San Luis Obispo. Trustees held up the award of a franchise to lay the line through the town until the Producers Transportation Company agreed to place a "tee" on the line in order to deliver oil to the city. The city officials argued the necessity of a "tee"—a pipe fitting which would make it possible to tap the line to remove oil—because, they said, if the line were going to pass through the streets of their city, it should supply anyone there with whatever oil they desired. The company agreed to install the "tee" and subsequently received the necessary approval to lay the line through the city.

Construction on the pipeline began on July 29, 1909, and the first oil began flowing through it to Port Harford in March, 1910. In one respect the new pipeline exceeded all expectations. Plans called for a 20,000-barrel-a-day capacity, but when the pumps actually began pushing, they proved able to drive more than 30,000 barrels of oil a day over the Coast Range to the sea.

Early operation of the line had its difficulties, Karge recalled. One bottleneck occurred during winters between Santa Margarita pump station and Avila. Because of the excessive distance between the two points, the incoming crude oil had to be heated at San Luis Obispo tank farm and helped on its way to Avila with booster pumps.

Because of the difficulty of moving heavy crude through the Kern River to Rio Bravo segment, not long after the line went into operation an additional pump station had to be built at Rosedale, six miles east of Rio Bravo. To further assist the task of moving heavy oil, the Producers Transportation Company pumped diesel distillate from Maltha Refinery at Bakersfield to Buttonwillow during the summer months, feeding it into the heavy crude during winter to lower its viscosity and facilitate its movement.

During construction of the pipeline a private telephone line that ran from pump station to pump station paralleled the pipe. Each pole was numbered, and Karge explained that "this became an excellent means of identifying the exact location of a leak." Pipeline walkers patrolled the line, equipped with horse-drawn vehicles carrying emergency supplies.

The men who patrolled the line detected leaky collars—couplings joining the individual lengths of pipe—or other sources of oil loss by telltale smudges on the surface. Where the leak could not be taken care of by the line walker, either by caulking a leaky joint or attaching a saddle to a worn or perforated section of pipe, he reported the leak to the dis-

The Tracy pump room, c.1912, was one of the installations from which the push was generated to keep heavy crude oil moving through the Associated pipeline. (Getty Oil Company)

patcher at San Luis Obispo, who sent out the nearest work crew to make repairs.

"Later the line-walkers made their rounds in automobiles," Karge commented, "which greatly expedited the process of leak detection and line repair, but was still slow compared with the low-flying airplane line 'walker' of today.

"In the northern pipeline office at San Louis Obispo, I recall that there hung a water color painting of the installation of the Producers pipeline," Karge said. "On the left of the picture, men with pick and shovel were excavating the trench; in the foreground others were connecting the threaded end of a twenty-foot section of pipe to a matching collar; on the right side, workers with the aid of horses and scrapers were busily filling in the trench and burying complete sections; but nowhere in the picture was there a string of unconnected pipe or a pile of pipe sections from which to draw! The source was either too distant to be included in the painting or the artist thought it was too trifling to be portrayed as a significant item. In any case, the missing pipe supply was regarded as something of a joke around the office," Karge said.

Karge also recounted an unusual coincidence connected with the Producers pipeline. A Scotsman named William Groundwater had worked as a machinist on the pipeline project, responsible for the installation of boilers, pumps and other equipment at each station. Later Groundwater became director of transportation for Union Oil with jurisdiction over the entire pipeline and tanker systems. He and Karge became well acquainted, working together on many projects.

On one occasion in the 1920's the two men travelled together to the division office at San Luis Obispo. The trip took them beside the Pacific Ocean on many stretches of the highway, and in the course of the journey, Groundwater began reminiscing about his boyhood days in the Orkney Islands, when he had sold his catch of fish to the crews of passing vessels.

The reminiscences struck a chord in Karge, who in earlier days had sailed before the mast and eventually graduated from wind-propelled vessels to the post of Third Officer of a German tanker, plying between New York and various European ports.

"In those days," Karge recalled, "tankers still burned coal under their boilers. Thus when leaving Europe for the United States, the vessel on which we sailed usually bunkered at North Shields on the east coast of England. Then we sailed north between the Orkney Islands and the Scottish mainland. As the ship made its way through the channels," Karge related, "we frequently came across small fishing boats and stopped while the fishermen drew alongside to trade their fish for Schnapps and tobacco."

In addition to his duties as Third Officer, Karge had the extra-curricular chore of officiating occasionally as purser to buy food for the crew.

Groundwater excitedly named some of the ships to which

Weber-Duller Co. of Los Angeles building a one-million-barrel concrete reservoir at the Producers pipeline tank farm near San Luis Obispo. (Huntington Library)

he had sold his catches. The tanker on which Karge had sailed was clear in his memory as one of them. The two men remembered the dates when they had been so occupied. The years coincided. Karge, it was plain, had undoubtedly been the party of the second part in the small business venture.

"Who would ever have thought that a German engineer and a Scottish fisherman," Karge exclaimed, "meeting in the tempestuous waters off the Orkney Islands, would subsequently be riding together along a smooth California highway in a luxurious limousine, piloted by a liveried chauffeur?"

In the years just after the turn of the century, Julius Fried sold groceries to oil people in the McKittrick, Taft, and Maricopa area. According to an account he related later to Graydon Oliver, a petroleum engineer, Fried hoped to acquire for himself a promising piece of oil land. At that time, drillers were the technologists of the industry, and Fried studied drilling operations, cultivating the friendship of the drillers, who were reluctant to divulge their trade secrets. Laymen could not easily tell from surface indications where production might be expected, but Fried finally won the sympathy of a successful driller willing to reveal his particular method of oil evaluation.

The driller took Fried off to the side of the derrick he was working on and, according to Oliver, said to him, "Look down there and you will see that the grass is red. The rainfall in the Taft-Maricopa area is very small but usually each winter there is enough moisture to grow small patches of grass that will head out in seedpods in the summer heat. In the intense sunshine, though, the grass will turn red. When you find that red grass, you have found oil land and probable production."

"With this secret and invaluable knowledge," Oliver related, "Julius Fried set out with his horse and buggy in search of the red grass. That was only possible when the summer was well advanced, and the seedpod had been well sunburned. By this means, he finally did locate a piece of land on which he filed under the placer mining laws and was granted a permit to do development work."

An apocryphal story? Perhaps, but beyond the locating of the land, one thing was certain. Fried lacked the capital to do more than erect the wooden shacks required to hold the property. He promptly set out to organize an oil company, primarily among his friends.

On December 9, 1908, the Lakeview Oil Company was incorporated under the laws of California, and with high hopes the company on January 1, 1909, spudded in to drill Lakeview No. 1.

The Lakeview gusher, three days after the wild well blew in. (Photograph by Robert B. Moran.)

With the derrick gone, the well spouted in a shallow crater, above. At right, men worked long hours building a sandbag levee to contain the flow. (William Rintoul)

Like more than one such venture, the Lakeview scheme proved considerably undercapitalized, and Fried had to return to his stockholders for more money. Work at the well dragged on. From the yellow clay, pebbles, and coarse sand the drillers had encountered at shallow depths, the cable tool bit passed into shales and heavy sand. A showing of gas came at 1,340 feet, but no oil was found. A neighboring company, Union Oil, offered help. Union, it developed, had its eye on the Lakeview land, not for its oil potential but as a site to build storage tanks for the Producers pipeline then being built.

The Lakeview partners made the best bargain they could. Union, though not overly enthusiastic, would continue to drill the Lakeview No. 1, now at 1,655 feet, but only on a spare time basis when Union crews might be available; in return, Union would get not only the right to place tanks on the Lakeview land but also control of the company—fifty-one percent of the capital stock and four of seven seats on the board of directors.

Drilling progressed slowly, and Lakeview stockholders grumbled that Union was dragging its feet. On the morning of March 15, 1910, fourteen months after the well had been begun, Walter Barnhart, the Union production superintendent, tethered his horse beside the wooden derrick and learned from the driller, Roy McMahon, that the bailer was stuck in the hole, then at a depth of 2,225 feet. Barnhart told McMahon to limber up the bailer by yanking the cable up and down. When the driller reversed power to take a strain on the cable, the heavy bailer blew out of the hole with enough force to send it crashing into the crown block at the top of the derrick. Oil spurted high into the overcast sky, men scattered, and Barnhart's horse bolted, not to be caught for a week. A classic blowout had begun.

The column of dark brown oil shooting up from the well could not be stopped. Flowing at an estimated rate of 18,000 barrels a day, oil demolished the wooden derrick. Sand buried the engine house, bunk houses, and coal shack. Spray from the well carried great distances. Housewives for miles around found their washing freckled with oil spots. Visitors parked their automobiles apparently out of reach of the spray, only to have the cars, and clothes as well, covered with petroleum mist in a few minutes.

News of the wild well reached Frank F. Hill, director of production for Union Oil Company in Santa Maria. He lost no time heading for the well, traveling by car through Cuyama Valley, not even stopping as he sometimes did to look at the oil seeps in the valley.

A torrent of oil, dubbed the "trout stream," poured away from the Lakeview gusher when Frank Hill took charge. The stream threatened not only to dissipate the oil so that it could never be recovered, but also to inundate Buena Vista Lake, the source of irrigation water for Miller & Lux farming operations.

Work began immediately on building huge earthen reservoirs to trap the oil in the sloping land between the well and

A river of oil from the gusher was named the "trout stream," below; at right, oil-soaked workmen work inside the gusher's crater. (William Rintoul)

the lake, eight miles away. All the teams and scrapers that could be hired in the Midway field, and some from as far away as Suisun City, 300 miles to the north, worked around the clock to build twenty huge sumps covering some sixty acres. The job cost more than $350,000.

Some 400 men labored to build a barricade around the well, lacing sand bags and sagebrush into a levee to hold back the flow of oil. Three pumps worked to full capacity delivering oil to a pair of 55,000-barrel tanks on Producers Transportation Company property at Maricopa. The tanks soon proved inadequate to handle the uncontrolled flow, which reached a peak estimated at 90,000 barrels a day.

Fire posed a constant threat. Frank Hill assigned experienced men to work with crews of newcomers, hoping to prevent the one mistake that could prove fatal. Guards kept sightseers at a safe distance. Fires were strictly forbidden in cookhouses on leases near the gusher. Activity at other wells in the area came to a halt.

The only casualties were the skin injuries suffered by many men as a result of the distillate baths required after ten- and twelve-hour shifts in the oily spray that enveloped the well.

Among the visitors who came to see the phenomenon were Gifford Pinchot, an adviser to Presidents Roosevelt and Taft, and his brother, Amos Pinchot. The gusher could soon be seen at night, for a well known as Tight Wad No. 3 on 25 Hill, four miles away, caught fire and lit up the countryside for miles around.

After eighteen months, the Lakeview gusher died, leaving a huge crater. (Robert B. Moran)

One man whom the gusher did not cheer was the city attorney of Bakersfield. He had acquired the Lakeview property in 1901 for five dollars and subsequently sold it for little more after a 1,000-foot dry hole was drilled on a nearby lease.

As oil flowed, the price of crude oil dropped to thirty cents a barrel. Standard Oil sized up the situation, built earthen reservoirs near Bakersfield, and bought all the cheap oil it could get, storing it for the day when the price would rise.

Finally, on September 9, 1911, 544 days after the well blew in, the Lakeview gusher caved at the bottom and died as suddenly as it was born. The secret of its production remained a mystery for many years, for deeper wells drilled a few hundred yards in each direction failed to produce. Eventually drillers learned that a very narrow bed of oil sand stood on edge in that vicinity. Only a few feet wide and more than a

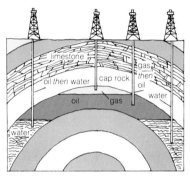

The diagram illustrates how oil may be found by some wells and missed by others drilled in the same field.

Sunset portion of the Midway-Sunset field, spreading eastward toward the Coast Range, in the years after the adventure of the nearby Lakeview gusher. (Getty Oil Company)

mile long, it ran at right angles to the other West Side sands. The Lakeview well was drilled about seventy-five feet outside the channel's edge, and the tremendous pressure in the underground reservoir forced the oil to push over into the Lakeview hole.

Lakeview No. 1 produced an estimated nine million barrels of oil, a record not yet equalled by any other well in California. The rescue operation saved more than four million barrels; the remainder was lost. Lakeview No. 1 remains the greatest gusher ever seen in California and perhaps the last oil well which successfully utilized red grass as a geological marker.

8 The Rotary Rig in California

In 1844, Queen Victoria issued the first patent for a rotary drilling rig to Robert Beart of Godmanchester, County of Huntington. In time, rotary tools would revolutionize the then unborn American oil industry, enabling men to drill holes deeper and faster. However, not until the 1890's did men apply the rotary rig to drill specifically for oil, completing the first rotary-drilled oil well at Corsicana, Texas, in 1895. Between 1895 and 1901, when the Lucas discovery well blew in at Spindletop, near Beaumont, Texas, oil men drilled more than 100 wells with rotary tools at Corsicana. At Spindletop, a rotary rig drilled the 1,020-foot well, which blew in for an estimated 100,000 barrels a day, establishing the state of Texas as a major oil source and earning widespread recognition for the rotary rig.

The basic principle of rotary drilling is the same as that for a carpenter's auger. A hole is made—in wood or the earth—by turning a cutting tool while weight is put on the tool.

On a rotary rig, a power source—initially it was a steam engine—furnishes power to turn the rotary table, which is located in the center of the derrick floor. The drill pipe goes through the center of the rotary table; when the table is turned, the drill pipe and the bit on the end of the pipe are turned, boring a hole as the pipe is lowered. As the hole is drilled deeper, lengths of drill pipe are added.

A circulating fluid, called drilling mud, is pumped down the drill pipe. The fluid passes through perforations in the bit, picks up rock cuttings made by the turning bit, and returns to the surface through the space between the drill pipe and the hole. At the surface, the mud travels through a screen, which removes the cuttings, and into a pit from where it is taken by the pumps and circulated back down the drill pipe to pick up more cuttings. Meantime, a mud cake becomes plastered against the wall of the hole. This cake prevents loss of the circulating fluid into porous rock and, together with the weight of the column of mud, prevents the hole from caving. The weight of the column of drilling mud also prevents

Rotary drilling rig, c. 1920
(*Huntington Library*)

high-pressure gas, oil, or salt water in a reservoir sand from blowing out of the hole when the sand is penetrated by the bit.

When a bit becomes dull, the drill pipe and bit are pulled from the hole. The job of hoisting pipe, and of lowering it, is handled by the drawworks—part of a block-and-tackle system with a number of pulleys on the crown and travelling block to increase pulling capability. As drill pipe is removed, it is stacked vertically in the derrick in stands. Each stand consists of a number of joints, usually two, three, or four, handled as a unit; the number of joints in the stand is determined by the derrick size. A new bit is then put on, drill pipe is lowered back into the hole, and drilling is resumed.

After Spindletop, the rotary rig began to come into its own. In 1907, 175 rotary rigs were operating in the Gulf Coast area. A complete rig then cost $2,825. A drilling crew consisted of five men, including a driller, derrickman, and three floor men, called "roughnecks." Men worked a twelve-hour tour; drillers were paid $5.50 a day; derrickmen, $4; roughnecks, $3. It took an average of 150 days to drill a 2,200-foot well, using only a daylight crew.

In 1908, Standard Oil decided to see if the rotary could do well in California. A company representative, J. R. McAllister, went to Louisiana in search of rotary rigs and drillers. He signed up six men to form the nucleus of crews to operate three rotary rigs.

One of the drillers was Lindsay (Lin) Little, from Gaffney, South Carolina, who five years before had quit his job as a fireman for the Southern Railroad to seek higher pay as a rotary roughneck in the booming Evangeline field in Louisiana. Little recalled that the rotary outfits were small and light. The power source was a nine-by-ten-foot single-cylinder steam engine. The mud pump was an eight-by-five-by-ten-foot unit. The line shaft, which was driven by a chain from the steam engine, was $2\frac{7}{8}$ inches in diameter; the shaft had a sprocket and chain for driving the rotary table. It also had a sprocket and chain driving down to the main drum shaft, which varied from $3\frac{1}{2}$ inches to $3\frac{7}{8}$ inches in diameter. The main drum shaft carried only the drum for spooling the drilling line, a brake flange, and a jaw clutch. On the brake

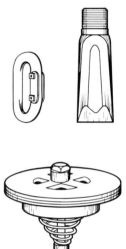

Rotary tools: C-link, drilling bit, and back-pressure valve.

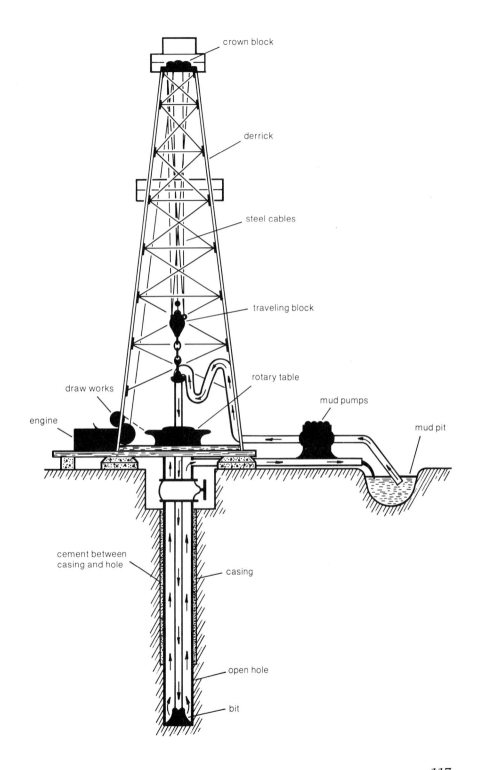

crown block

derrick

steel cables

traveling block

rotary table

mud pumps

mud pit

draw works

engine

cement between
casing and hole

casing

open hole

bit

The rotary rig.

117

Drawing of the Bell rotary table, invented by Cy Bell in 1910. (Standard Oil Company of California)

flange, the brake bands were three to four inches wide. There were no brake blocks. The brakes had metal-to-metal contact and, naturally, the steel dust caused considerable irritation.

The rotary table used a grip-ring drive; that is, it contained a system of "slips" (pieces of steel of wedge-shaped design) that permitted the table to grip the drill pipe. This was necessary for turning the pipe in a way that gained cutting action from the drilling bit on the end. The mechanism did not work very well, however, because it caused grooves along the pipe passing through the grip. Eventually the grip-ring drive became obsolete and was replaced.

The rotary crew from Louisiana arrived in California to work for Standard Oil on August 17, 1908. In the crew were Lin Little, Roland Little, Cyrus Bell, Neil J. Norris, and John Rodoffer. Lin Little recalled that they met a cool reception in some quarters. Ordinary human resistance to change combined with the anxiety of cable tool drillers who feared the loss of their jobs. The latter referred to their rotary counterparts as "swivelnecks" and looked on them as interlopers. Some operators, moreover, considered the rotary method unsuited to California conditions. The operators argued that the rotary mud necessary to wash cuttings to the surface and to line the walls of the hole in lieu of the pipe normally carried on down during cable drilling would seal off the oil sand and do irreparable damage to wells.

Others took a more hopeful approach. "I thought the rotary could be made into something that would work," recalled Earl Delaney, who began his career as a cable tool driller in his native West Virginia, "but it certainly was a crude affair when I first saw it around 1909." The rotary rig had no guard, Delaney pointed out, "the chains would break and fly around all over the rig, and you never knew when you were going to get socked with something."

Some companies began to experiment with rotaries. W.K. Oil, of which J.W. Pauson was secretary, tried out near Coalinga in 1902 what may have been the first rotary drilling rig in California. "After drilling to 2,400 feet," said Pauson, "the hole was so crooked that we could not run the casing. We changed to standard [cable] tools, straightened the hole, completed the well, and condemned the rotary as not fit for California use!" More successfully, Amalgamated Oil had drilled a hole in the vicinity of the Salt Lake field (Los Angeles) to a depth of 2,357 feet in only five weeks, an amazing performance, but unfortunately no oil was found.

Standard represented the most substantial commitment yet to the new tool. The company announced plans to drill the first rotary hole on the Talara property on Section 24, T32S-R23E in Kern County's Midway field.

To quell undue optimism, or perhaps to dispel future criticism if the method did not work, a Standard spokesman said the company planned a cautious approach and did not expect to complete the first well in less than four months. The company announced another innovation for the rotary well. The well would be completed with "screen." Instead of using slotted pipe as a liner opposite the oil-producing sand to hold the sand in place, Standard would use screen similar to that used by fire engines in straining water pumped from ditches.

While Standard officials publicly said they saw no reason why the rotary rig would not work on the West Side, others lacked such confidence. Associated Oil Company was trying out a rotary on the Vernon lease on Section 29, T28S-R28E in the Kern River field. After making less than fifty feet in ten days, or the incredibly slow footage rate of less than five feet a day, the company said it considered the rotary method a

failure, at least in that particular field. Boulder beds encountered at shallow depth created a problem. While the up-and-down action of cable tools had pounded such "cobbles" out of the way, the rotary turned around on top of them without making appreciable headway.

While one of Standard's three rotary rigs began work on the Talara property, the rig to which Lin Little was assigned went to the Salinas Valley, where Standard had leased 20,000 acres on both sides of the railroad and the Salinas River between San Ardo and Bradley. Before the well could be spudded, the management decided against drilling the hole and moved the rig once more to Altamont Pass between Oakland and Tracy, where drilling began in October, 1908. Standard had taken extensive leases in the area, announcing plans to spend not less than $50,000 for a thorough evaluation. Along with the rotary rig, the company assigned a cable tool rig to drill another "wildcat" or exploratory well two miles away. Inevitably, an element of competition crept into the undertaking, pitting the rotary drill against cable tools. Three months after drilling began, the rotary was down to 600 feet; the cable tool rig was at 700 feet.

The winter was a wet one, and drilling was frequently interrupted by heavy rains that made access to the rigs over existing farm roads all but impossible. Both jobs dragged on into 1909, and they both ultimately turned out to be dry holes.

Meanwhile, on the Talara property in the Midway field, Standard had reached a depth of only 1,720 feet more than four months after spudding in, making progress slower than normal for a cable tool rig. Cobbles blocked the rotary's advance, as they had blocked the Associated experiment at Kern River. The rotary could not drive them out and only succeeded in washing away the softer material from between the rocks, which then fell together, massing directly in the path of the drill bit and collecting in the basin created by the hydraulic stream. Also, Midway lacked a suitable clay for use in mixing drilling fluid to circulate in the hole.

There were, however, glimmers of success with the rotary drill. In Los Angeles, Amalgamated announced that the rigs were practical for drilling to depths of 1,500 to 2,000 feet, at which point it would be prudent to replace them with cable tools known to be reliable at such depths. At Santa Maria, Associated reported that after working three months to reach a depth of 450 feet in a cable tool hole, the company had installed a rotary and in less than two months had reached a depth of 1,100 feet, passing through sand, gravel and clay beds to solid blue shale, using only two strings of casing.

California Oil World reported that the coming rig would combine rotary for rapid work through clays and loose formations and cable tools for hard formations. The magazine quoted an unidentified supply man who suggested a parallel with operating an automobile. "You run your machine over level unobstructed ground on the high speed with everything wide open," he explained, "but when you come to climbing a hill or driving over a furrowed road or cobblestones, you throw in the low."

Equipment people announced the production of more powerful rotaries. The Johnson rotary manufacturers said they intended to make a tool that would cut through granite at the rate of one foot an hour. Standard Oil announced plans to test several new drilling rigs, stating that so far it appeared that the rotary rig was slower than the cable tool method, but that the rotary method achieved a saving in the cost of pipe. With rotary drilling, the drilling mud preserved longer sections of open hole than were possible with cable tools. If the drill bit did not find oil, it would not be necessary to run pipe. Savings also came about with successful wells because the rotary hole did not need to be as large in diameter and could be cased with smaller pipe.

While oil men debated the merits of the rotary, the next stop for Lin Little and the crew that drilled the dry hole at Altamont Pass was the town of Taft, only recently known as Moron, on the booming West Side. Little recalled that the day driller was the boss of the rig. The regular working day of the drilling crew was twelve hours, and there were two shifts, day and night. The day driller hired his own night driller—Little hired George Kenniston as his drilling partner—and the rest of the crew and was responsible for bringing in the well. He took samples from the mud ditch, and not infrequently brought in wells, simply by his sense of smell, without ever catching sight of an oil sand.

It took from sixty to one hundred days to drill a 3,000 foot well with rotary tools. The drilling fluid, Little recalled, ran from ten to fifteen percent sand most of the time; the sand shaker, a device to separate sand from fluid, had not yet been invented; the only sand shakers available were the rough-necks themselves, who separated sand from drilling fluid with long-handled shovels, removing sand and shale before fluid circulated back down the hole.

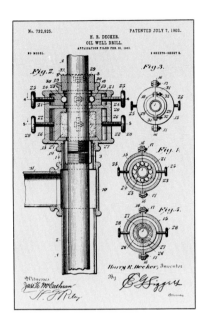

Patent Office sketch for the Decker Blowout-preventer, 1903. (American Petroleum Institute)

For drill pipe, crews used the same type of pipe that was used for pipelines. The low-carbon, lap-weld steel pipe broke off in the threads; it twisted in two; and the seams parted under pressure. The fishing jobs that ensued to recover the lost pipe cost valuable time. Nearly every fishing job meant a "washover"—encircling the stuck pipe with a socket pipe so that it might be freed from the hole—or a "side track"—to get around the fish. Sophisticated fishing tools had not been invented, nor had the technique of circulating crude oil down the hole to free stuck pipe come into use.

In the area where Standard was drilling, Little recalled, there was a gas cap—a reservoir containing more gas than the oil below it could dissolve—that lay at comparatively shallow depths ranging from 2,350 to 2,700 feet. When wells were shut in, the pressures on surface equipment were as high as 1,800 to 2,000 pounds per square inch. Blowout preventers, the equipment designed to prevent a well from blowing in out of control, were far from efficient, and blowouts were frequent when the drill bit penetrated the gas cap. Also, operators largely depended on horse and mule teams to haul supplies to rigs. This did not speed up operations when a crew was waiting for essential tools.

The technique employed in the beginning, Little recalled, used rotary tools to drill the hole to the point at which casing was to be landed. Then they changed the rig over to cable tools to run the casing. A float valve had not yet been developed, and the pipe would fill up on the inside with mud and cuttings. The cable tool crew would bail it out, then make a few feet of small hole for the purpose of shutting off water. As an alternative, the crew occasionally dumped cement into the bottom of the hole with a dump bailer. Then they pulled the casing up twenty to thirty feet and filled it with water, screwed a solid plug into the top of the casing and let it down, forcing the cement out around the shoe of the casing. After cementing, a well usually had to stand about three weeks while the cement hardened.

Though the going was not easy, Lin Little and other rotary drillers made progress. Perhaps the biggest breakthrough occurred in January, 1910, when Standard Oil brought in a

The twin steam engine, 1910, that revolutionized rotary drilling. (Standard Oil Company of California)

gusher on Section 30, T32S-R24E in the Sunset portion of the Midway-Sunset field. The well flowed 1,500 barrels a day of 21°-gravity oil from 2,432 feet. It was the first gusher in the state to be drilled by rotary tools.

While the completion of good wells helped, another factor hastened the transition from cable tools to rotary in California. Cable tool drillers, who at first feared they would lose their jobs, quickly learned the new techniques and became rotary drillers themselves.

One of the biggest contributions to rotary drilling came from Union Tool in Los Angeles with the development in 1910 of the twin drilling engine. Dick Smith recalled that "we had a great deal of trouble with the single engine which would stop on center, making it necessary for one of the crew to go back and kick it over, which was very dangerous." To start the engine, the crew man would have to crank it, as motorists had to do with early cars, and, like the motorists, the crew man ran the risk of injury if the engine happened to backfire. In the case of the crew man, the size of the engine made the risk greater.

The idea of using two engines in tandem to avoid the problem occurred to Jim Pickering, field superintendent for Union Oil at Brea. Pickering noted the operation of double-cylinder donkey engines on docks, and decided that if two single-cylinder drilling engines were placed parallel to each other and the two crankshafts replaced with one long crank-shaft with two cranks on it, offset from each other by 90°, it would be impossible for the crankshaft ever to stop on dead-center.

"Jim brought in two single-cylinder drilling engines and gave us an order to do the mechanical work and furnish new parts where necessary," Smith recalled, "and when the job was completed, he took the new engine out and put it to work." It proved satisfactory, and Union Tool went on to develop and market the twin-engine drilling machine.

Though Pickering had the original idea, he was too busy to get it patented, and its immediately widespread use prevented him from ever getting a patent. Smith recalled that Pickering never profited from his idea.

As rotaries became more common, oil operators and man-ufacturers consulted, and the latter began to build bigger and better equipment. In 1911, Union Tool built two of the first rotaries fabricated in California, delivering them to Standard Oil for use at Taft.

Improvements came fast. Frank Hill, the Union Oil super-intendent who directed the battle to contain the Lakeview gusher, built scale models and carried on extensive investiga-tion of the factors affecting the course of the rotary drill bit. As a result, he was one of the first to advocate the use of heavy drill collars—the fitting which attaches the bit to the drill pipe—in order to hold the entire drilling string in better sus-pension and to insure a straighter hole. Before that, recalled Earl Delaney, "we used to give the bit weight by lowering the weight of the drill pipe. Now they put the weight on the drill collars and keep the pipe in tension. That makes for drilling straighter holes and prevents twist-offs and all those things."

Hill also was one of the first to sense the relationship between rotary speed and the weight on the drilling bit and invented a weight indicator to enable him to control the relationship.

To stop pipe failures, manufacturers tried heat treating. This helped, as did the later development of careful control in processing high-carbon steel. In the early 1920's, seamless drill pipe appeared and ended most of the failures common to lap-weld pipe. Alloying increased the strength of pipe.

Gas engines began to replace steam on rotaries in the 1920's. In spite of its rough action, compared to steam, the gas engine was lighter and cheaper to operate than steam equipment. When the depression of the 1930's demanded low cost and portable equipment, gas-powered rigs replaced more and more steam rigs. As wells went deeper, imposing a greater strain on derricks, steel began to replace wood, completely replacing it in the 1930's.

Improved drilling mud made it possible to drill deeper and deeper. At Spindletop in 1901, drillers drove cattle through their surface mud pits, thickening the water with clay to better hold back loose surface sands. Later, drillers tried to improve their mud by adding to it clays obtained from surface outcroppings. In the 1920's, a Louisianan decided that if he could make the mud heavier, it would ward against blowouts. Initially, he added iron oxide to the mud. Later he added barite, a common mineral of high specific gravity, which quickly became the most commonly used mud-weighting material.

"Of course the most important thing was the kind of mud they were using," Earl Delaney pointed out. "We used to use whatever mud was made by drilling—it was sometimes ninety percent sand. If you twisted off, you had a washover job," he said. Sometimes the mud would get very bad, Delaney reminisced. "I worked on a rotary once, and we got down 700 or 800 feet and got a bunch of boulders washed out in the hole and there was no mud—nothing to pulverize those hard boulders. We fought that thing for weeks," Delaney said. "The contractor pretty nearly went broke, and I left the job feeling sorry for him."

Problems inherent in the early use of drilling muds, not immediately recognized, were high filtration water loss and

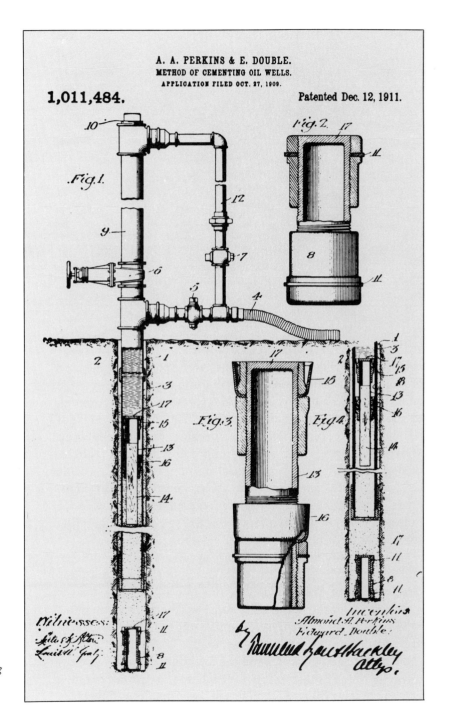

Patent Office sketch for the Perkins-Double Two-plug cementing method, 1909–1911. (American Petroleum Institute)

Earl Hampton, center, with tools used in early Ventura well. (Earl Hampton) At right, the new rock bits that improved rotary drilling. (American Petroleum Institute)

excessive solids content. Water lost to the formation because of filtration caused shales to swell and slough into the hole. Mud solids left a filter cake on sand formations thick enough to make the drill pipe stick. As a result, fishing jobs and loss of the hole were common. Areas became famous for their "heaving shales" which seemed impossible to drill. In 1928, drillers tried the first commercial bentonite in a California well. Not actually intended for use in an oil well, the high-grade clay had been mixed and processed as an additive for Portland cement. The bentonite helped to control sloughing formations and cut down on filter-cake thickness.

In the 1930's mud engineers first appeared in the oil fields and soon began to play an important role in the systematic development of deep drilling.

At the rotary table, the kelly replaced the grip ring assembly. The kelly was a heavy square pipe; the rotary table gripped it through bushings set in the table. As the table rotated, it turned the kelly and the drill pipe extending below the kelly to the bottom of the hole where the drill bit cut new hole. The kelly slipped through the square hole in the rotary

table as footage was ground out. When it was necessary to add more drill pipe, the bushings were pulled out and slips were set to hold the drill string. The drilling crew unthreaded the connection between the kelly and the drill pipe, added another joint of drill pipe, made up the connection between the kelly and the pipe, and continued drilling ahead.

New fishing tools took advantage of the new rig's powers of rotation and circulation. A new overshot allowed mud to circulate through the fish and imparted torque that helped to free it. An outside cutter encircled the fish and enabled the crew to retrieve it by cutting it up into pieces instead of either pulling it out in one piece or trying to break it off at connections. Small-diameter cutting tools enabled crews to cut stuck pipe from the inside, making retrieval even easier.

For twenty years after the rotary came to California, the old fishtail bit—a holdover from cable tool drilling—remained the dominant drill bit. The fishtail bit had only two cutting edges. These dulled quickly, and drillers tried making bits with three and four cutting edges. This did not help very much until manufacturers began putting tungsten carbide and other hard materials on the cutting edges. After that improvement in the early 1920's, bits cut better in both hard and soft formation drilling.

The "drag" bits with their gouging action on the bottom of the hole put a great deal of torque and shock on the drill string, especially when too much weight was run on the bit or when a big hole was drilled with small pipe. The pipe failures which resulted gave impetus to development of the rolling-cutter rock bit, which, as the name implies, featured rolling cutters rather than a sharpened surface on the face of the bit. By the 1920's, the rock bits had enabled the rotary for the first time to drill faster than cable tools in hard-rock country, and the rotary rig became the dominant one in the California oil fields.

9 Geologists and Engineers

Along with alterations in oil production technology went changes in the training of oil men. A new breed of college-trained, professional geologists and engineers began to enter the industry, and gradually their contributions came to be an essential part of the oil production process. At first, however, the new professionals found themselves unappreciated and underpaid.

Walter English, a prominent California geologist who worked for Standard and Superior oil companies for many years, received one of the first master's degrees in paleontology awarded by the University of California. "When I got out of school in 1913," he recounted, "Doc Merriam, who had taught me most of my stuff, sent a recommendation for me to the U.S. Geological Survey, and they hired me at $75 a month. We had a teamster and a cook who got $80 in our same outfit—more than the geologists were paid."

A few years earlier, after getting his B.A. in geology, English had looked around for a job. In Los Angeles he spoke with Bill Orcutt, the vice-president in charge of exploration for Union Oil.

"Mr. Orcutt," English said, "how about giving me a job as a geologist?"

Orcutt replied politely, "I'm sorry I can't give you a job. Union Oil Company already has a geologist. As a matter of fact, I am a geologist myself, so we are completely supplied with geologists."

"That seemed to be the general outlook of many of the oil companies," English recounted. "The exploration plan they were working under was—in the first place to get their land on seepages, and in the second place they tried to get close to someone who had discovered something." English pointed out that seismic and gravity surveys were just beginning. "I can remember," he said, "that geologists were not in terrific demand although you would think that in the beginning of the business they would have been crazy about having geologists to locate stuff for them. But they weren't," English said.

Geologists Ralph Arnold, left, and H. R. Johnson, in field outfits, near McKittrick, 1908. Arnold carries a camera, binoculars, and a field bag for rock samples. Johnson adds a canteen. (Huntington Library)

131

Earl Gaylord, who worked as the chief petroleum engineer for Standard Oil, started out as the only college student working as an oil field laborer for Kern Trading and Oil Company, near Maricopa in the San Joaquin Valley. After graduating from college in 1911 with a degree in engineering, he found work as a geologist for the geological department of the Southern Pacific Company, whose director, E. T. Dumble, also headed up K. T. & O.. "At first I made some maps in the office and then I was put on land classification work in the foothills of the Sierra Nevada," Gaylord recalled. Then he became part of a mapping party for the west side of the San Joaquin Valley, working from what is known as the McDonald anticline down to the Grapevine at the southern end of the valley. "We mapped some very interesting country, mostly on foot or horseback," Gaylord said. He was transferred for a time to San Francisco before being located at Kerto as "a resident geologist, there being no such title as 'petroleum engineer' in those days," he said. "When I moved to Kerto," Gaylord pointed out, "that $100 a month meant a pretty good salary because, as I recall, I paid $10 or $15 a month rent for my house—with services, including ice!" Not until 1926, when Standard Oil absorbed Pacific Oil, K. T. & O.'s successor, did Gaylord acquire the title of petroleum engineer.

When Roy McLaughlin, who had first worked as a geologist in the gold mining industry, went to work for Associated Oil in the San Joaquin Valley, he travelled about in a light wagon with an assistant who served as cook and camp helper. "No living quarters were provided for us at the main company lease," McLaughlin recalled, and he and his wife rented rooms in "such small houses as were available." Most of the drilling crews and their families lived on the lease in "what was called 'Rag Town,' a collection of tent houses."

"I was given no definite instructions," McLaughlin recalled, "other than to collect well logs and histories, draw cross-sections and construct underground contour maps." The chores did afford an excellent opportunity to observe drilling operations "which were carried on with tools that showed little improvement over those used by Colonel Drake when he drilled the first oil well in the United States."

State Mining Bureau format for drawing well logs, 1916.

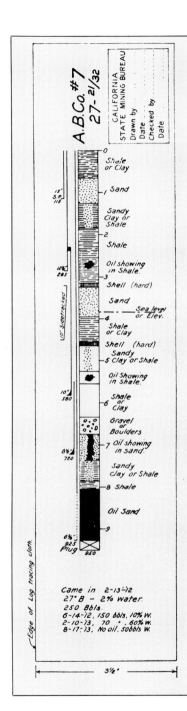

Left column figure labels (well log):

A.B.Co.#7
27- 2/32

CALIFORNIA
STATE MINING BUREAU
Drawn by
Date
Checked by
Date

0 — Shale or Clay
1 — Sand
15' S.P. 115
Sandy Clay or Shale
2 — Shale
12½" 285
Oil showing in Shale.
3 — Shell (hard)
Sand
Sea level or Elev.
4 — Shale or Clay
Shell (hard)
Sandy Clay or Shale
5
Oil Showing in Shale.
10' 580
Shale or Clay
6
Gravel or Boulders
7 — Oil showing in sand.
8½" 720
Sandy Clay or Shale
8 — Shale
Oil Sand
9
6⅝" 925
Plug 950

10" Sidetracked

Edge of Log tracing cloth

Came in 2-13-12
27° B - 2% water.
250 Bbls.
6-14-12, 150 bbls, 10% W.
2-10-13, 70 " , 60% W.
8-17-13, No oil, 50bbls W.

3½"

CALIFORNIA STATE MINING BUREAU.

FLETCHER HAMILTON, State Mineralogist.

DEPARTMENT OF PETROLEUM AND GAS

R.P.McLaughlin, State Oil and Gas Supervisor.

CONVENTIONAL SYMBOLS
FOR
MAPS AND WELL LOGS
July 1st, 1916.

WELL LOGS

Formations as shown in typical drawing. (clay or shale, and sandy shale, 2 symbols)
Cement as shown on 10" and 8½" casings.
Formation shut-off as on 12½" and S.P.
Perforations as on 6⅝" casings.
Adapter as shown between 10" and 12½"
Casing cut and pulled, as on 8½" casing.
Casing sidetracked, as on left, 320' to 450'
Casing shot, collapsed, split, or otherwise altered, should be noted on left margin.
Gas, Water and Oil should be noted at right of log.

MAP SYMBOLS – WELLS.

o   Rig in place.
⊕   "   abandoned.
◐   Incompleted
    "   and   "
●   Completed
◆   "   and   "
Y   Water
✶   "
☼   Gas
✳   "

--- Note ---

Graphic logs of wells will be drawn on strips of tracing cloth 3½ inches wide, without margin lines. Scale of 100 feet to one inch.
General arrangement and spacing as on example herewith: (top, title: righthand, geologic data: lefthand, casing record and mechanical data: bottom, producing conditions).
The tracing may be folded and filed with the written log.

Cross-sections of several logs will be made by fastening individual tracings, in their proper relative positions, on a larger piece of tracing cloth, using gummed stickers at top and bottom. The drawings may also be pinned directly to blue print paper before placing it in the frame. The only title on the cross-section will be large figures (about one inch in height) in the upper right hand corner of the drawing, indicating the section, township, and range – thus: 27 $\frac{21\ S.}{32\ E.}$

Cross-sectional blue prints will be filed in an ordinary letter cabinet, being folded so that title numbers are visible without unfolding the drawing.

As an aid to quick and uniform drafting, a guide beneath the tracing should have the scale and necessary vertical lines.

Symbols are simplified so that they can be made with a right line pen at the same time that dividing lines between formations are made. Ordinarily, the symbol alone will suffice, without name of formation.

W.E. CONDON

Later, McLaughlin was transferred to Taft. "I was provided with an automobile of an early model, a Locomobile, which frequently developed failures at unexpected places, requiring the driver to walk several miles for mechanical repairs. The duties of my job were still vague and such recommendations as I made, suggesting the acquisition of land, seemed to produce no results. Even after receiving a considerable increase in salary, I concluded that the job had no future." McLaughlin's resignation in favor of a San Francisco consultantship and an eventual position with the State Mining Bureau's new Oil and Gas Division came about a year before the company abolished its geological department.

As one of the first white-collar professionals in the field for Standard Oil, Ross McCollum also found the going rough. In 1917, he said, "there were only three college graduates working in the producing department in Southern California. One was a boiler inspector, one was a safety engineer, and myself. During the next few years I attempted to develop many new

U.S. Geological Survey party in the Devil's Den district of Kern County, south of Dudley, July 1908. (Huntington Library)

ideas, but because of the feeling that existed between the technical men and the field men, practically none of my recommendations were accepted."

The burgeoning oil industry, geared to drilling by seeps and punching down holes, was initially slow to recognize changing times. Not long after the college-trained men began to enter the industry, however, its leaders found that a company could overlook their contributions only on pain of being left behind by competitors.

Walter English, who had gone to work for the U.S.G.S. at a pay less than cooks or teamsters, continued with the Survey. He spent about five months out of each year in the field and the remainder of the time in Washington, D.C., writing up field reports, which were published in technical bulletins with maps attached. "There was nothing fancy about them," English recalled, but these reports, and others he later prepared as an independent consultant and for Standard Oil, helped identify new California oil locales.

Others of the new breed made important contributions, too. Charley Scharpenberg, who had left Illinois to work for Standard in California, devoted considerable time to the development of improved cementing practices for shutting off water in wells, and he became an expert on rotary mud as well. Ross McCollum also saw acceptance of his ideas, advanced years before, for chemical additives for quick-setting cement in water shut-offs.

The acceptance of college-trained men had its impact in the colleges. Harold M. Van Clief recalled that "in the fall of 1919, Dr. Bailey Willis, head of the geology department at Stanford, quoted oil production consumption and reserves data to his class in structural geology and wound up by telling them, 'The world is at your feet.' " Van Clief said that "reaction at the student level was immediate." Enrollment in the 1920 field geology class totaled sixty-one students, an all-time record number.

Van Clief and half of his classmates ultimately entered the oil industry. Included among the group were the following men: Art Abrahams, who went with Richfield Oil; Harry Abrams, North American; Pedro Aguerevere, Venezuelan

Seepage brings oil to the surface when not blocked by impervious rock.

Ralph Arnold at work, Santa Barbara County, September, 1906. Of the picture Arnold wrote: "Near view of flinty Monterey (Middle Miocene) shales on Sisquoc River 6 ½ mi. northeast of Sisquoc . . . ." (Huntington Library)

government; Dave Anderson, Schlumberger; Karl Arleth, Standard Oil; Lloyd Aubert, General Petroleum; Glen Bowes, Ohio Oil; Harold Brown, Pacific Oil; Douglas Bundy, Whittier Associates; Austin Cable, Standard Oil; Willard Classen, Tidewater; George Collins, Standard Oil; Ralph Copley, Standard Oil; Gene Davis, Signal Oil; Dwight Deckman, Cities Service; H.L. (Speed) Driver, Standard Oil; Chet Gibbs, Standard Oil; Bart Gillespie, C.C.M.O.; F. W. (Pretz) Hertel, Tidewater; Ike Holston, Tidewater; Cecil Housch, independent operator; Bill Kleinpell, Union Oil; George Know, Standard Oil; H. M. Oliver, Hancock; J. W. Paulson, Petroleum Securities; Lawrence Porter, Richfield; Perry Roberts, Honolulu Oil; Al Sands, Texaco; Ted Sawyer, Schlumberger; Dick Sherman, Barnsdall; Rouse Simmons, Bolsa Chica; L. M. Spencer, Midway Gas; Max Steineke, Standard Oil; T. F. Stipp, U.S. Geological Survey; Harry Stolz, consultant; Don Weaver, Wilshire Oil; R. G. (Peg) Whealton, Standard Oil; K. A. Wright, Barkus & Wright; and Myron Zandmer, independent operator.

The importance of the new professionally trained men can be seen in their work in developing the Ventura oil field.

In April, 1916, Ralph Lloyd, a trained geologist whose father had been a cattle rancher in the Ventura district, submitted to Shell Oil a large acreage block along what was known as the Ventura anticline.

Lloyd had every reason to believe there might be oil. As a youth more than twenty years before, he had helped his father run cattle there. Once, in a violent storm, young Lloyd's father had saddled up and gone out to look after the cattle. The events that followed, related by Ralph Lloyd to Graydon Oliver, a consulting petroleum engineer, left an indelible impression. When Lloyd's father reached the top of the ridge where the Pacific Gasoline Company's plant later would be built, a flash of lightning struck the ground, starting a brush fire. Flames leapt toward the elder Lloyd, who rode to what looked like a barren patch of ground. Flames burned there too, apparently fed by escaping natural gas. Lloyd escaped into a gorge, but his horse burned to death.

Unable to forget the phenomenon, Ralph Lloyd felt the need of some geological understanding and persuaded his father to send him to the University of California at Berkeley. There he became intrigued by the similarity between diagrams that were used to illustrate the anticlinal theory of the accumulation of oil and the structural characteristics, as he knew them, of the Ventura Avenue area. However, after only two years at the university, Ralph Lloyd was obliged to return home when his father became ill and needed him to help manage the ranch.

Aided by a friend who was a surveyor and geologist, Lloyd, on his return, mapped the Ventura Avenue anticline and

became more convinced than ever that it held a rich oil repository.

The elder Lloyd, in need of cash, sold the ranch to sheepmen, whose faith in the area's oil potential did not match Ralph Lloyd's. The purchasers of the ranch were not willing to pay an additional $5,000 for the mineral rights, and the Lloyd family retained them through Ventura Land & Water Company, a family corporation.

When he mapped the Ventura Avenue area, Ralph Lloyd deduced that the area covered an extensive structural trap. Based on this conviction, he leased other adjacent parcels, such as the Taylor, Hartman, McGonigle and Gosnell properties. Hartman, a rancher, had once dug a water well which tasted salty. He put a cover over the well, and a few days later went back, removed the cover, and lit a match to look inside for the cause of the smell. Fortunately, he was not injured by the ensuing gas explosion.

After Lloyd had leased the various properties, he tried to

Ventura Avenue in 1914 before the drilling boom, above (Ventura County Pioneer Museum); at right, portion of a diagram showing the axes of Ventura Avenue anticline. The Taylor, Lloyd, and Hartman properties appear across the top; Ventura Avenue runs vertically in the middle of the diagram.

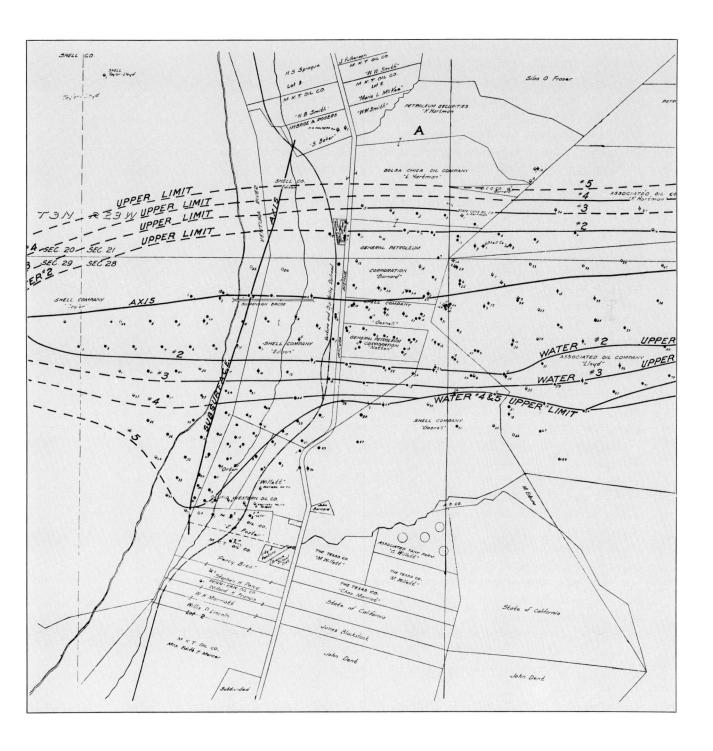

interest operators in the development of the land and approached Shell Oil. Several years before, the company had hired as perhaps its first American geologist, John E. (Brick) Elliott, a Stanford graduate who had "geologized" the area in the summer of 1910 for Associated Oil. He had submitted a report and map of the area to Stanford in lieu of regular requirements for graduation, and by virtue of this work the Shell geological staff was already alerted to the general characteristics of the Lloyd property.

Shell signed a contract for the acreage block, after which the firm's geological department assembled all its forces and started a geological investigation, possibly the first detailed geological survey of a wide area completed in California. Every dip taken in the field was "shot in" with a transit and rod. In that way, the company obtained both the exact location and elevation. The final subsurface topographic map developed from this data was projected to a depth of 5,000 feet, deeper than any producing oil measures then known; in the succeeding years of exploration and development, the map proved to be very accurate.

Even before submitting the large acreage block to Shell, Ralph Lloyd with Joseph Dabney and E. J. Miley had started a well on the huge structure along the east bank of the Ventura River. This well "blew out" during the fall of 1916—after Shell had taken the acreage block submitted to it—and produced gas under high pressure associated with 56°-gravity oil. The gravity of the oil was so high that some car owners clandestinely filled their gasoline tanks with it. The blowout proved that the large structure would yield high gravity oil under high wet gas pressure. It also demonstrated that developing the new pay would not be easy.

"After four years of bitter disappointment and the expenditure of several hundred thousand dollars," Ralph Lloyd later wrote "we had nothing to show but craters blown out of the earth by gas, strings of wrecked casing deep in the earth, and a few puddles of oily salt water. The gas and the salt water seemed to overpower any efforts of man and machinery to get at the oil."

Shell faced a great problem in deciding the proper method

Wooden derricks occupied steep hillside locations in the early development of the Ventura field. (Earl Hampton)

for developing the high gas pressure area at Ventura. The use of cable tools permitted a sampling of the formation being drilled, but offered no way of either controlling the gas pressure or preventing "freezing" of casing caused by caving formations. The latter required circulation of mud fluid as used by the rotary system of drilling. In order to prevent gas

141

blowouts and to obtain samples of the formations, Shell's engineers decided to use the so-called circulating cable tool method. This proved very expensive and not very satisfactory.

Never had drillers hit anything like Ventura. Drilling would proceed only a few feet before a gas pocket at shallow depth would be encountered. If drillers were lucky, the gas would only blow out the string of drilling tools. If they were not so lucky, the tools, casing, and rig would go vaulting into the air, and a geyser of water and gas would blow for a few days. Only after patiently rebuilding the rig could they get on with the job.

The continual loss of mud circulation presented a steady hazard. A sudden drop in mud pressure usually meant that high-pressure gas had been encountered, and the mud soon turned to froth. If heavier mud were not pumped down at once, a blowout would ensue. Sometimes the mud did not encounter gas but would go wandering off through an underground fissure created by one of the field's many cross-faults. Once, after pumping mud frantically for hours, a crew discovered a small stream of mud issuing from a surface outcrop a quarter-mile away.

For every few feet of hole the men succeeded in drilling, they lost hours and days in other endeavors, usually mixing and circulating mud or in trying to set casing against high-pressure water. One of Shell's drilling crew members, Charles Hansen, was working on a company well on the McGonigle lease when he left to go into the army in 1917. The United States entered the World War and helped to win it. Hansen was discharged and returned to find his crew still drilling on the same well, not much deeper than when he had left.

The well on the McGonigle lease cost more than any other exploratory well that Shell drilled on the acreage block acquired from Ralph Lloyd. The tract of land lay far back from the road and was separated by a high ridge from any convenient roads. The only way to get to the well site appeared to be to build a long, expensive road. Instead, Shell built an aerial tramway a mile and an eighth long over the rocky ridge to connect the drill site with the road on the nearest lease. A heavy steam boiler to provide power could not be transported

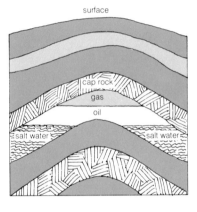

Impervious rock traps oil in a subterranean reservoir.

Government survey parties became a familiar sight in California oil fields. This camp was at the Santa Fe Railway headquarters in the Midway district, Kern County, 1908. (Huntington Library)

because the cable could not carry that much weight. Instead, electricity provided the power, and this also eliminated the problem of hauling fuel. The McGonigle well was one of the first in California to be drilled entirely by electricity.

In four years of drilling at Ventura, Shell laid out about two and one-half million dollars, big money in a day when most oil wells cost only a few thousand dollars to drill. For the large outlay, Shell had only one discovery. In February, 1919, the company completed Gosnell No. 1, getting an initial production of 150 barrels a day of 29°-gravity oil, cutting seventy-five

Mountainous terrain posed problems for surveyors putting in well locations at Ventura. (Getty Oil Company)

percent water, from an interval at 3,200–3,498 feet. Total depth of the well was 3,498 feet.

To those less optimistic than Ben van der Linden, Shell's general manager in California, it began to look as if Ventura were a poor bargain. In October, 1919, the company abandoned the unsuccessful McGonigle well and late that same year, Shell surrendered the McGonigle lease.

Others in the area did not do much better. In December, 1919, three companies had wells in the Ventura field: Shell, General Petroleum, and State Consolidated Oil. During that month, the field's eight producing wells yielded a total of 425 barrels of 50 + °-gravity oil, 3,678 barrels of 28°-gravity oil, eleven million cubic feet of gas—and 100,000 barrels of water. Efforts to shut off water had proved futile. Of approximately fifty attempts to get water shut-offs, only two had been definite successes. Three were regarded as probably successful, and about forty-five had been pronounced failures. Action of the gas prevented cement from setting. The heavy mud used to kill gas settled around the shoe and caused cement to channel up one side of the hole, leaving the other side open to infiltrating water. Loose porous formations, moreover, allowed water to make its way around the cement and into the hole below the shoe.

Meanwhile Wilhelm van Holst Pellekaan had come from The Hague to liquidate Shell's role in the field. He began by suggesting that Brick Elliott alter his original affirmative report. In view of all the facts demonstrated about the structure and the presence of oil and gas, Elliott refused to do this. Pellekaan next offered to sell the leased area east of the Ventura River to Associated Oil. Cort Decius, chief geologist for Associated, promptly accepted the offer. Pellekaan then attempted to sell Shell's interest in the remainder of the leased area west of the Ventura River. At this point, Ben van der Linden interposed to prevent further dispersals, which proved a fortunate measure given the ultimate richness of the area.

Ralph Lloyd still endeavored to justify his faith in the Ventura field, but the area remained difficult to drill. Not until Associated Oil, in the person of Pretz Hertel, drilled a well on the Lloyd property and managed a successful water shut-off, did Ventura Avenue become generally recognized as a proming field. Shell, too, later enjoyed success in drilling the field.

The Ventura story would be incomplete without reference to Mrs. Grubb, owner of the Taylor properties. Graydon Oliver related that she had leased her holdings to her former schoolmate, Ralph Lloyd. Shell Oil tapped the Taylor lease for considerable quantities of oil and sent royalty checks regularly to Mrs. Grubb. She did not deposit these checks, however, and Shell sent a representative to ascertain why. She admitted receiving checks, but apparently never had any occasion to use them, so she had just held on to them. For safety's sake, the representative urged her to deposit the checks in a small local bank nearby. By now, the checks amounted to almost $500,000, and the manager of the bank was reluctant to take them. Shell solved the problem, Oliver explained, "by getting one of the larger banks to take over the little bank and accept Mrs. Grubb's checks."

With the discovery of the Signal Hill oil field in Long Beach in 1921, the California oil industry passed another milestone in its saga of productivity. The opening of Signal Hill also shows how rapidly the process and technique of oil exploration had changed in the two decades since the Elwood brothers dug out the discovery well at Kern River with their own hands.

As early as 1918, Frank Hayes, a Shell geologist working under the direction of Brick Elliott, had recommended that leasing and exploration be undertaken at Signal Hill, a highly promising structure in the southern part of the Los Angeles Basin lying within the city of Long Beach. Frank Vaughan, a colleague of Hayes who had recently joined the company, recalled that the Signal Hill proposal did not meet with overwhelming enthusiasm. "Unfortunately," he said, "very important changes in the organization of the Shell geological department had taken place. Wilhelm van Holst Pellekaan had been put in complete charge and had entered upon a program of harshly criticizing Elliott, including Elliott's recommendation to drill at Ventura Avenue. When Hayes recommended Signal Hill, van Holst promptly condemned it. He pointed out that four holes had been drilled in the nearby Dominguez Hills structure without favorable results."

Hayes countered with the argument that "Signal Hill was a much larger structure and that it seemed safe to assume that in the earlier stages of the structural development of both structures, most of the oil had migrated to Signal Hill." Over the ensuing two days of discussion, Hayes brought up this matter from time to time. "Hayes was a mild mannered man,"— Vaughan recalled, "not at all the kind to be provoked into active combat—but he was persistent. Finally van Holst said, 'Hayes, you are a nice fellow and I like you and I know you mean well, but in this matter you are absolutely wrong. Let's hear no more of it.' So Signal Hill was dropped."

Van Holst went with a number of the staff to projects outside the state and left in charge Alvin Theodore Schwennesen, a 1911 graduate of Stanford University. "He was not left

Signal Hill, 1930. (Science Collection, Department of Geography, U.C.L.A.)

147

in California to carry on active exploration, but it was the feeling that some geologists ought to be on hand just in case anything happened," Vaughan said. Schwennesen busied himself looking over what had been done, and in the spring of 1920 he ran across Hayes' report on Signal Hill. In April he employed Dwight Thornburg to check over Hayes' map and report on Signal Hill. Favorable comments by Thornburg led Schwennesen to check personally some of the critical areas indicated by Thornburg. He concluded that a good prospect existed and, in Vaughan's words, decided that "certainly something ought to be done about looking into it further."

The decision whether or not to drill was an agonizing one for Shell. They had a five-year losing streak in California and had spent three million dollars in the thus far unsuccessful attempt to bring in commercial production at Ventura. And only four years before, Union Oil had drilled an unsuccessful well on Signal Hill. The company decided to take the risk, however, and Shell began to lease land. The process was a difficult one because Signal Hill was developing as an exclusive residential area and some expensive homes had already been built. The company spent $60,000 to lease 240 acres.

Others had their sights set on Signal Hill, too. Although Shell managed to lease the Bixby Land Company property, the company did not succeed in acquiring all of the available leases on top of Signal Hill. Joe Jensen, geologist and land agent for Amalgamated Oil, had an opportunity to acquire 240 acres but, of course, he first had to convince his management in San Francisco. In December, 1920, one of the top officials came down with him to Signal Hill and Jensen suggested, "Right over that gravel pit I'd drill my first well." The executive said he had been over all the country and had visited most of the territory down to Newport Beach, and there simply was no saturation. He cancelled Jensen's negotiations for the 240–acre parcel.

For the site of its first hole, Shell selected a location below the brow of the hill on the edge of a tract belonging to Alamitos Land Company. The program, in accordance with practice then current, called for the Alamitos No. 1 to be drilled with rotary tools to a depth at which good oil showings

Cross-section of the Signal Hill field prepared by the State Oil and Gas Supervisor, 1923.

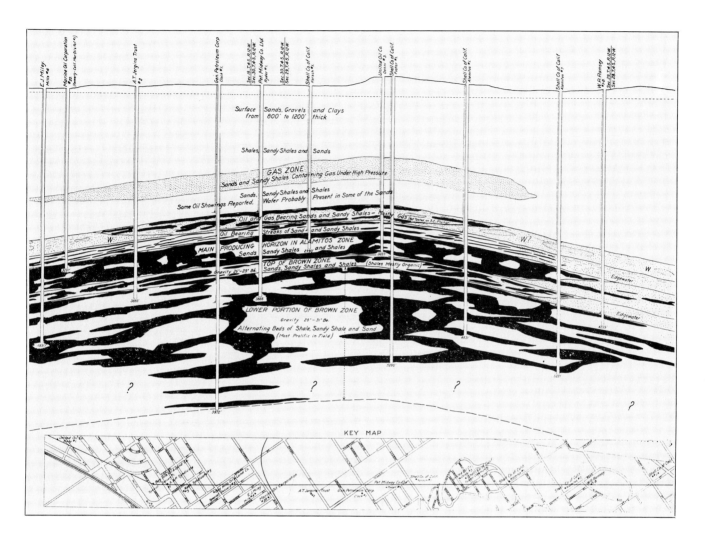

were obtained. Then, to avoid blocking off the oil sands with rotary mud, the well would be finished with cable tools.

On March 23, 1921, a Shell rotary crew from Oilfields, near Coalinga, moved in and started drilling. By May 2, they had reached 2,765 feet. A core sample revealed oil sand. Encouraged, Shell raised $50,000 to lease more land. Crews ran casing to the bottom of the hole, and a cable tool crew arrived to finish the well. While making a water shut-off test, the crew found seventy feet of oil standing in the hole, along with a strong showing of gas. The good news traveled fast, and until the company built a substantial barricade, the crew had to spend much of its time keeping sightseers off the derrick floor.

Drilling had reached a depth of 3,114 feet by June 23 when oil blew over the crown block 114 feet in the air. Then, the well sanded-up and stopped flowing. For two days, crews worked to get the hole cleaned out. Excitement ran high, and when the clean-out job was finished and oil from Alamitos No. 1 started flowing to tanks at four o'clock in the morning on June 25, 1921, 500 spectators were on hand to see the sight. During its first twenty-four hours, the well made 590 barrels of 22°-gravity oil, valued at $1.20 a barrel. Soon production increased to 1,200 barrels daily.

"Signal Hill is the scene of feverish activity, of an endless caravan of automobiles coming and going, of hustle and bustle, of a glow of optimism," *California Oil World* reported. "Derricks are being erected as fast as timber reaches the ground. New companies are coming into being overnight. Every available piece of acreage on and about Signal Hill is being signed up."

At the time the well came in, Frank Vaughan recalled, van Holst was in Ecuador, showing little interest in California. "As soon as the well was in, he was back promptly," Vaughan said. "The whole activity of Shell Oil assumed a new slant. Shell's Alamitos No. 1, the discovery well of Signal Hill, was a milestone in the history of oil exploration on the Pacific Coast."

Vaughan credited Schwennesen with the discovery. "I know the kind of a fight he had to put up to get the thing across," Vaughan emphasized. "It's true that Hayes was the first man to map it. It's true that Thornburg was called in, but that was done by Schwennesen. He might have called in someone else. The man who really had to take it up with the authorities to get things done—and who did it in spite of some opposition—was Schwennesen."

The find promised to further enhance California's situation as an oil supplier, the state being the only one in the nation, according to American Petroleum Institute figures, with a surplus of production over consumption. The A.P.I. reported

In May, 1928, Richfield's well at Signal Hill, at right, was the deepest producing oil well in the world, getting 4,100 barrels a day from 7,409 feet. (Huntington Library)

that the United States had used some 53.5 million barrels of oil in March, 1921, of which 40.8 million barrels represented domestic production and 12.7 million barrels imported from Mexico. The industry group said that if there were no imports from Mexico, the country would be short some 116,000 barrels per day. The Signal Hill discovery came at a time when California produced more oil than any other state, putting out 337,000 barrels daily, the bulk of it from San Joaquin Valley fields. Exports from the port of Los Angeles ran around $1.1 million worth of petroleum products each month,

Boilers furnished the power for drilling operations, and crews rushed to rig them so that new wells could be started. (Division of Oil and Gas, Long Beach)

mainly to Chile, which took more than $600,000 a month worth of gasoline and fuel oil, illuminants, and lubricants. Lesser dollar values of the same products went to Canada, the Philippine Islands, Nicaragua, El Salvador, Mexico, Japan, Honduras, Colombia, and New Zealand.

The discovery at Signal Hill naturally spurred an intensive leasing campaign. Property values soared, and land agents reportedly paid as high as $20,000 an acre for oil and gas rights, prompting *California Oil World* to editorialize, "It may be a fine thing for the landowners, but likely to be tough on the stockholders of the company paying such a preposterous sum."

Though Joe Jensen had been thwarted in his first efforts to get his company to take land, he was not willing to give up. The surface dips on the west side of Signal Hill were very steep and discouraged leasing activity west of Cherry Avenue, where the homes of the wealthier people were located. Jensen rented an amusement hall in the vicinity and invited all the adjacent lot owners to discuss the question of leasing. At this meeting he advised them to make up a community lease and also suggested that the time for making the best deal had not yet come. He asked them to make a good community lease and keep it intact, and as a reward for his own efforts, to give him the first chance at the lease when it was finally ready. This they did. But when the landowners asked for a one-fourth royalty, all of Jensen's superiors in San Francisco turned the deal down. Union Oil, however, took the lease. It became known as the Union Community lease, and on it was drilled a well which ultimately produced some three million barrels of oil.

Meanwhile, Shell Oil lost no time following up its discovery, putting six drilling outfits to work within a matter of weeks. The company purchased the home of Andrew Palii, a retired Frenchman who several years before had chosen the top of Signal Hill, which overlooked miles of country in one direction and the Pacific Ocean in another, as an ideal home location. Shell turned the mansion into bachelor quarters for employees and expended every effort to develop quickly the newly found oil field. The company's second well to hit pay sand blew in a gasser and caught fire, burning an estimated twenty million cubic feet per day. The company rigged a battery of seventeen boilers to pour steam on the well, and with steam and mud subdued the wild well in a thirty-six-hour battle, but not before the fire had destroyed the drilling rig.

In quick order, the company lost two more rigs to blazing wells. At one, the Martin No. 1, the well was standing cemented at 2,640 feet when it got away. Shell brought in two fire-fighters, Ford Alexander and G. E. Oliver from Taft. Wearing asbestos suits furnished by Johns-Manville, they set up a stand a few feet from the wild well, placed a hundred-pound charge of dynamite on the stand, connected the charge by electrical wire with a detonator seventy-five feet back, and proceeded to snuff the flames seventeen hours after they had begun.

An intensive drilling boom on town lots followed the discovery. The steam issuing from the boilers that drove drilling rigs shrouded Signal Hill in fog on windless days. The number of rigs working in the field rose to 270, so closely spaced a man could walk from the floor of one rig to another without setting foot on the ground. Ten months after the discovery, 108 producing wells put out 14,000 barrels daily, divided among thirty-seven companies. One 4,000-barrel-a-day well stood above the gravel pit that Joe Jensen had unsuccessfully tried to lease.

The production from the booming field proved to be the first installment on what would climb in succeeding years to more than one-half million barrels of oil per acre, making the field one of the richest in terms of per-acre production that the world has ever seen. In the wake of the find, the Long Beach Women's Club held a public discussion to debate the pros and cons of turning the city's business district into an oil field. The city manager said if the wells were big enough, it might pay to replace business blocks and residences with derricks and refineries. However, he hastily added, they should be assured the wells would be big, and the city should not be in a hurry to tear down city buildings.

For Shell, the splendid find could not have come at a more opportune moment. It gave the company, hitherto unrepresented in Southern California, the local production it needed to enable it to build a refinery and begin the retail sale of gasoline and other petroleum products in the West Coast's largest consuming area.

From Signal Hill, the boom spread to Santa Fe Springs, ten miles away. Clifford Davis, a driller working for Petroleum Midway Oil in the fields close to Richfield Station on the Santa Fe lines near Fullerton, recalled how the word came to dismantle rigs and boilers to get ready to move to booming Santa Fe Springs. "We hauled the machinery with old Federal trucks, Mack trucks, Morelands, Whites, Brockways, large teams of horses and mules and some by railroad—any way to get it there in a hurry," Davis said. Wooden derricks were springing up like trees. Large piles of lumber one day, a derrick in two or three days; men sawing with large crosscut saws, sawing the timbers up by hand to form the various parts

From Signal Hill the play spread to Santa Fe Springs, above, where more than one well got away. (Union Oil Company of California)

of the derricks. Only the best lumber was used. It came by trainloads, then was hauled by teams and trucks. The sound of rig hatchets rang throughout the day, and when lights were available, they rang out in the night air.

The first well Davis worked on was near the Pacific Day Products plant at Los Nietos. While he was working on the midnight shift, a well south of his dramatically blew out, shooting large columns of mud and gas into the air. Davis said that "the Mexicans and people of Santa Fe Springs had their things loaded into buggies, on horses, into wagons, Model T Fords and moved out of the townsite. Houses were dark; the noise was deafening. With the loud roar was a large stream of mud and water running in all directions, soon forming a large mound. It blew for weeks and weeks. As fear lessened, the people moved back."

The boom pace continued, and Davis described how operators crowded derricks onto small lots, "putting the boilers in the street if you didn't have enough room on your lot. Sometimes the boilers was between two tin-can shacks, as many men built their living shacks in the street. Some had a wife and two kids in them, but mostly just men. One or two was women's 'business houses.' The noise was bad, but business was good," he said.

"It was everybody for himself," Davis explained. "There was a lot of drilling contractors moving in to drill wells for stock-selling companies and promoters or anyone who wanted a well drilled and had the cash to pay for it.

"Things had moved along in the boom to where the gamblers was coming in and the magazine-selling girls. They got our money but we never got no magazines," Davis recalled. Sometimes unscrupulous salesmen had their problems. "We would look out from the rig floor and see a roughneck chasing another one around the block with a rig hatchet in his hand trying to catch up with the guy to part his hair. One of my roughnecks came running and said hadn't we better stop them? I said, 'Hell, no. One more pass around the block and they will both be winded.' "

Davis had a drilling crew composed of "one grocery clerk, one dry goods clerk, one cop, one college student. If one quit, you could get another in five minutes," he said. "But you never knew what you were getting as these guys came from all parts of the United States," he pointed out. "Some were nice and some were rough and tough. The rough and tough ones were the ones you had to watch for. They would do almost anything. Brass knuckles, a knife or a hatchet, a crowbar.

"It was said that someone sent posters East and South and said there was a shortage of men and that the pay was twenty to thirty dollars a day, which was not true," Davis explained. "Large companies at this time paid drillers $9 per day. Derrickmen and roughnecks got $7.50 per day. The contractors would pay $1 to $1.50 more a day."

A friend of Davis' "had a service station in Fullerton. He told me of people stopping to get directions to Santa Fe Springs where they heard from posters you made twenty to thirty dollars a day. He told me of a family in a Model T Ford touring car with two small children, a wife, a gas stove, a tent and a chicken coop fastened on each side to keep a goat for fresh milk for the kids. Wherever they stopped, there was all the comforts of home."

Santa Fe Springs, April 1930.
(Division of Oil and Gas, Long Beach)

Some of the newcomers took up residence on Norwalk Boulevard between Santa Fe Springs and Norwalk. The area was known variously as The Gum Grove, Springs Slums and Kings Camp. In the grove were "girls in tents with a hot bottle of soda pop. These girls had an ironing board and a shirt to iron. They kept ironing that same shirt while waiting for business customers," Davis observed.

Gamblers moved in. "They made buildings of one-by-twelve boards with bolts known as California-style buildings," Davis explained, "mostly with front and back rooms, always with a quick escape back door. In the front would be pool tables and a tobacco counter; in the back, card tables and, of course, slot machines—five cents, ten cents, twenty-five cents and one dollar. The pool tables were seldom used to play pool on, mostly to bank the dice on. All gambling in these places was done with poker chips which you could buy from the man running the tobacco counter or owner redeemable with a twenty percent discount for the house. Some of these buildings were two-story with rooms to rent upstairs with one bath to a building. They built these gambling buildings all over Santa Fe Springs, Norwalk and Los Nietos."

Fire at Getty's No. 17, Santa Fe Springs, 1928. (Standard Oil Company of California)

The most noted of all the gambling places was the Old Baker Winery Barn. Davis remembered that it was "all right to visit these gambling places in the daytime alone but in the night we went in two's or the whole crew would go. These gambling places were run all day and most of the night. The hijacks would clean them out every so often. They would take the safes and break the slot machines for what money was in them. The hijacking business got stronger as the boom went on."

Davis' boss, a six-and-a-half-footer who weighed 220 pounds and was known as "Slim" Travis, was presiding over a gambling game, Davis recalled, "in the head office of the drilling company when some guy opened the door and said, 'Stick 'em up.' Slim Travis hit him. Two shots went through the roof. Slim had a cut on the side of his head, could have been by a bullet, and the man got up and ran as Slim's blow had knocked him down a four-foot flight of steps. Someone told him, 'Slim, you damned fool, someone could have been shot.' Slim said, 'Yes, and we could have been out two thousand bucks.' "

Amazon Drilling Company crew, Santa Fe Springs, 1922. The photograph was taken by an itinerant photographer who set up his tripod on the rig floor and promised the crewmen a suitably mounted picture for fifty cents apiece. Guy Miller, donor of this photograph, is pictured at right. He was the derrickman on the crew.

Sandwich and coffee stands proliferated, and their counters had three-inch backboards on them for rolling dice. "Blanket gamblers," Davis said, "were all over the place. When trouble would start, they grabbed blanket, money, dice and all and ran." Honky-tonks were everywhere, and in the eucalyptus groves tents and tin buildings housed brothels and drinking places.

"When we was working afternoon tour and off at midnight we used to make the tour of these places," Davis recalled. "Los Nietos, Gypsy Town, Baker's Barn and the Gum Grove. My crew and me was transferred to the Howard Bell lease between Butler Road and the Santa Fe Railroad track. It was here that we became acquainted with May's place, a real up-to-date place. Pool rooms, gambling, eats and all. The gambling parlors was in the back and rooms upstairs and a restaurant which hired eight waitresses." Davis said that "after you hung around this place long enough and they knew you, then May would put a pin button on you. It said, 'May-day.' Now you had the badge, you could come in anytime as they knew you were all right. If you wanted to, you could give

Oil field restaurant, Signal Hill. (Garth Young)

your pay check to May and this entitled you to everything for a month—eats, gambling, girls, and all. After midnight, the tables in May's place was used for shooting craps. Anywhere from fifty to one hundred men was in there and it went until four o'clock in the morning."

Another time Davis and his crewmates went to a party in the old Baker Winery barn. "Well, we got there," Davis said, "and they had a stage show going. The girls were as bare as the walls. It was like going to a carnival. Once you got in you could see the girls shimmy for nothing, that is if you didn't shoot craps or play the blackjack table or draw poker or roulette wheels or slot machines and you was not out anything. It was about one-thirty when we was in one end of the building watching some hula dancers when a police whistle began blowing. We all figured how we was going to get out. Just then somebody pulled the light switch and most everybody got away okay but the boys that run the place. There was girls with evening dresses on and guys with tuxedos climbing over the fences and running in all directions." Davis said that "we learned that night that it was better to leave your car and walk a half mile than to get caught in a mess like that."

Davis explained that "the country was supposed to be dry but Scotch whiskey was plentiful as was most hard liquors. They was plenty of bootleggers and blind pigs around. Old Jim Doty could supply anything you wanted. You left your money by a eucalyptus tree that night and went back to dig in the sand for your bottles. The bootleggers sold the booze at so much a kick. The real strong was six kicks, or $6. They made a drink on what is now known as Florence Avenue that was made from persimmons and pineapple called 'one way.' One glass and you were down. Old Kentuck—or Tennessee—ran the place and made the stuff in the Fred Humpler barn.

"Them was the great days of the boom," concluded Davis.

11 Safer and Shorter Working Days

Early on a January evening in 1922, the crew on afternoon tour at Union Oil's Alexander No. 1 at Santa Fe Springs was at 2,060 feet when mud began to boil out of the casing. The driller shouted at the men to run. There was a man up the derrick. He started to climb down. When he was about thirty-five feet from the ground, a column of mud and rocks shot out of the hole, rising to the crown block. The derrickman jumped. Fortunately, he landed in the sump. Though shaken, he escaped with his life. Drill pipe blew out of the hole, which cratered, swallowing up the drilling rig and the two automobiles the men had driven to work. Within hours, the site of the well was a cauldron filled with swirling mud and debris.

On a winter night at Signal Hill, cold foggy air settled over the field. A crew working morning tour at a well on the Anticline Oil Syndicate property went into the changeroom to warm up. One of the men struck a match to ignite the open gas burner. Six men were seriously burned in the explosion that followed. Fog hovering over the field had held down gas, allowing a pocket to gather in the changeroom. Sometimes fog held down gas until pockets accumulated under viaducts. A spark from a passing car might cause an explosion, endangering the lives of the occupants of the car as well as destroying the machine.

At Huntington Beach, a sand line snapped at Jameson Oil Company's No. 2. The wire line, suddenly released from tension, whipped like a scythe over the rig floor, killing one man, seriously injuring another. Another crew member was so unnerved he was taken to a hospital in Santa Ana.

The accidents were part of a sad picture, the casualty statistics of work in California's oil fields. Even as the state's production rose to more than 100 million barrels a year, the highest it had ever been, the accident rate rose. H. C. Miller, associate petroleum engineer in the San Francisco office of the Department of Interior's Bureau of Mines, tabulated a total of 4,109 accidents during 1921–1922 in the producing departments of ten representative oil companies operating in Cali-

Capping a wild well, Ventura, c. 1920.
(Getty Oil Company)

fornia. The state's Industrial Accident Commission listed ninety-eight fatalities and 406 men permanently injured.

Why? Sometimes men were killed because boilers blew up. Steam furnished power for drilling, and some boilers were poorly made. Many were poorly operated. None had automatic water-level or fire controls. When the fireman saw a low water level, he had a natural tendency to run in more feed-water. Cold water hitting overheated tubes too often caused an explosion. During 1922, the Bureau of Mines recorded thirty-four oil field boiler explosions. Of these, twenty-seven were caused by low water. Few of the boilers exploded without taking human life.

The early rotary rigs had no clutch for the chain-drive. As a result, the rotary chain ran all the time that the drilling engine ran, and this caused numerous accidents when members of the crew got tangled up in the rotary chain. The first crude attempt to prevent such accidents consisted of nailing up a two-by-eight-inch board so that it lay between the upper and lower sides of the chain. Since the board was more easily seen than the chain, this helped keep men out of the chain. The board had the added advantage of keeping the tight and slack sides of the chain separated in case it loosened. Later, the chain guard made the drilling floor safe for the crew.

Sometimes when the driller was running pipe back in the hole, he had a tendency to throw on the brakes to check the drill string's rapid descent. The intense heat generated on the brake-flange rims produced a stress that expanded the rims, weakening drum flanges so much that many flanges burst, throwing particles fanwise over the derrick floor. H. C. Miller of the Bureau of Mines noted in his tabulation of accidents that such bursting flanges had been responsible for six recent fatalities.

Many of the traveling blocks used in rotary drilling had no guards. Since the derrickman often had to grab the lines to twist the blocks about, it was easy for him to have his fingers caught between the lines and the sheaves.

To compound the safety problem, the drilling boom in Southern California oil fields brought both a rush to get wells drilled and inexperienced men to the rig floor. Contractors

Fighting fire at Santa Fe Springs, 1920.
(Union Oil Company of California)

began to take over the drilling of wells, and the pressure mounted to drill holes faster and faster. Initially, the operator generally paid the contractor a stipulated sum at the start of the job, another sum when the first 1,000 feet was reached, another at 2,000 feet, and the balance when the well was finished. To keep money coming in, the contractor had to make hole as fast as possible. By early 1923, one contractor was drilling twenty-six wells at Long Beach, ten at Santa Fe Springs, four at Huntington Beach and one each at Torrance and in Ventura County. The fields were alive with talk of fast drilling times, including the claim by a crew at Dominguez of a world's speed record of 3,250 feet of hole in fifteen days, or an average of 239 feet a day, with the best time being 290 feet in one tour.

Gaslight illumination was provided on this cable-tool drilling rig from a "yellow dog," center right. Such lamps were common on early drilling rigs, and were a safety hazard. (Crooks Stafford, Petroleum Production Pioneers Collection, Long Beach Public Library)

Under such boom conditions, little thought was given to safety. Clifford Davis recalled that some contractors were known "for killing so many men from drilling lines breaking, hoist drums blowing up and dropping pipe or getting caught in the open chains, as there were no safety things. They were rushing their men nearly to death to get the holes bored, so they could get on to another one to cash in on the big money. Losing men was no problem as the streets was full of them who had rushed here for the boom," said Davis.

The high accident rate drew a response from state and federal governments. The state's Industrial Accident Commission met with oil operators to solicit safety proposals and early in 1924 called a public hearing in Los Angeles to consider the resulting tentative safety orders.

These tentative orders called for guards on moving parts

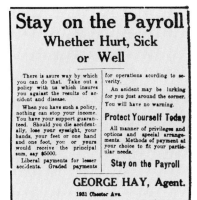
(*Beale Memorial Library, Bakersfield*)

such as the chain and gears of drawworks or shaft-driven rotaries, sheave-wheel guards for traveling blocks, an auxiliary means of escape for the derrickman by a specially-rigged life line or by a guy line smooth enough to allow him to slide on it to the ground, and safety belts for those men working above the level of the rig floor. The orders would require that any cellar sixteen feet or deeper have a runway or stairway to the surface, and that if the cellar—which accommodated the blowout prevention equipment—were less than sixteen feet deep, it should have a ladder. Also, the orders specified that a first aid kit be provided at each drilling rig and that all accidents be reported.

On the federal level, H. C. Miller interviewed field superintendents and workers on the subject of safety and photographed installations, particularly drawworks, engines and rotary tables, in order to make them safer. Miller also undertook a classification of more than 4,000 accidents under 103 subtitles indicating such data as extent and disability, part of body affected, and days lost. He used this information in a paper on oil field safety. The effort by Miller was only part of the Bureau of Mines' approach, which included exhibiting through oil regions a film, "When Wages Stop," illustrating the dangers of oil field work.

Oil companies mounted safety campaigns. In San Francisco, Roy W. Kelly, manager of Associated Oil's industrial relations department, helped bring a group of companies together to make a safety film, which the companies screened for groups of employees. A second step, Kelly said, called for the preparation of safety handbooks to be put in the hands of superintendents, foremen, and workmen.

In Los Angeles, the Chamber of Commerce conducted eight safety schools over a two-month period, awarding a silken honor banner to Standard Oil for the attendance of more than one hundred of the company's employees at each of eight meetings.

At Redondo Beach, Union Oil introduced a new-style ladder for its rigs. The ladder was guarded by wire cage to protect the derrickman in case of slippage when he climbed or descended from his working board to the derrick.

Tool companies joined the safety campaign. Union Tool provided the bolts of crown block bearings with an extra safety nut to prevent an operator from dropping a bolt while making an adjustment. The company introduced steel sheaves to eliminate the danger of badly worn cast iron breaking and scattering pieces of metal on men below. The company also introduced a latch so that the openings of hoisting hooks could be completely closed. The firm also pledged to eliminate sharp corners and projections wherever possible in designing machinery. Edward Timbs, a Union Tool executive, said, "In making any safety device we find it somewhat difficult to convince buyers that there is enough value in the device to warrant the expense but, as a result of this great educational campaign for safety, we hope some time to have every piece of drilling equipment fully guarded when it leaves the factory."

On March 1, 1924, the Petroleum Industry Safety Orders, compiled by the state's Industrial Accident Commission with the cooperation of oil operators, went into effect. California was the first state to adopt such orders.

Improvements in living conditions in the company field camps and pumping stations had preceded the concern for improving safety measures. The living quarters provided at the more isolated outposts could be ample indeed, with cottages for married employees and bunkhouses for the single men. A journalist in 1908 wrote that at the camp operated by California Oilfields Ltd. at Coalinga, the bunkhouse had "every modern convenience, bathrooms, sinks, running water, and other pleasurable luxuries." Many camps also had recreation halls, where movies were shown regularly.

Companies prided themselves on the food they served in the camp cookhouses. At Honolulu Oil's main camp in the Buena Vista Hills near Taft, Mrs. Pearl Yahl cooked for almost thirty-five years. When she first went to work there, the company had a garden and also kept chickens and a half-dozen cows. "I used to fix all kinds of things like baked heart and tongue and liver," she recalled. Later, tastes changed. "The men wanted chops and other cuts of meat. I used to fix a lot of chicken and dumplings. Later, all they wanted was fried chicken." Tastes in desserts changed, too. "One of the big

Burning Hartnell sump to destroy unwanted accumulation of waste oil sent a cloud of black smoke over Union Oil's camp at Orcutt, c. 1904. (Robert B. Moran) (Below:) Fire drill at Fellows, 1910. (William Rintoul)

Honolulu's main camp mess hall.
(Warren Kraft)

events in the old days was when we froze ice cream in the old ten-gallon freezer the company used to have. Later when everyone had ice cream in the refrigerator at home, they demanded homemade pie," Mrs. Yahl remembered.

Working conditions also improved as the movement for the eight-hour day gained adherents. From the beginning of the industry, crews had worked twelve-hour shifts, day or night. In 1917 Standard and several smaller companies went over to the three-shift, eight-hour day, although the seven-day work week remained customary for drilling crews until the late 1920's. At Standard, other shift employees changed to the six-day week in 1919.

The Oil and Gas Well Workers Union gained a great deal of strength in the California oil fields in 1917–18 in an effort to extend the shorter working day and a minimum daily wage of $4 to all companies. These demands were endorsed by a federal mediation commission seeking to avert an oil strike during wartime. The operators had little choice but to go along. With the end of the wartime conditions, however, the California oil companies no longer felt the necessity of negotiating wages, hours, and working conditions with unions and the federal government. In July, 1921, a conference of oil operators meeting in Los Angeles voted unanimously that they would not renew their agreements with the union, now renamed the International Association of Oil Fields, Gas Well

Victims of rig accidents could be treated at Oil Fields Emergency Hospital, Venice. (Title Insurance and Trust Company)

and Refinery Workers, A.F.L., when the pact expired at the end of August. In addition, the operators said, they would cut wages $1 a day because of the "decreased cost of living."

The union reluctantly accepted the wage cut, which for lower categories of workers amounted to a seventeen percent cut, but announced it would strike to prevent further reductions. The union demanded that operators agree to a memorandum of terms, with the federal government as a party to the agreement, that would prohibit further wage cuts for one year as well as other changes in working conditions. Some union men feared a return to the twelve-hour day.

The shutdown came at midnight on September 11, 1921, when about 8,000 San Joaquin Valley oil workers struck some 425 companies, affecting the West Side oil fields as well as the Kern River and Coalinga fields. In Kern County, the strike affected some 213,000 b/d production, most of it on the West Side.

Three days after the strike began, the first skirmish occurred at Pentland Junction near Maricopa. More than 2,000 strikers and sympathizers turned back a Southern Pacific

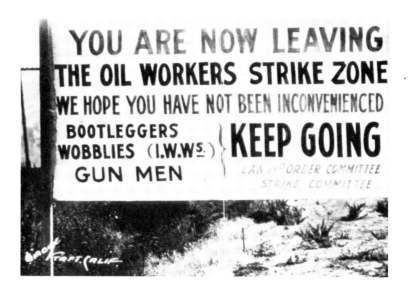

(Left:) Striking oil workers take to the streets in Taft, 1921. The strikers were not hospitable to certain types specified on their sign (right). (Kern County Museum)

special carrying men from San Francisco to the oil fields. The union called them strikebreakers; operators said they were guards. Sheriff D. B. Newell promptly banned the sale of firearms and ammunition in Kern County.

The union, proclaiming it would maintain law and order, combed its ranks for ex-servicemen, organized them into patrols identified by red, white and blue badges, and promptly expelled several I.W.W. organizers. Operators hotly claimed the patrols were in fact pickets. The union let it be known that it would tolerate no foolishness. A union spokesman said: "If any of the boys on strike secure a drink of liquor, the committee [Law and Order Committee] will find out where it was secured and the place will be raided by police operatives." Bootleggers took the hint and went on vacation.

Clifford Davis recalled that one of the roughnecks on his crew belonged to the I.W.W. "He was burnt up in one of these gambling buildings called the Arcade—fifty rooms— on the General Petroleum Jalk lease on Norwalk Boulevard. It burned so fast that nobody seen anything," Davis said. "After a short investigation, it was just one of those things of the past. The Sheriff asked me what I knew about him. I told them all I knew about him was that he was an I.W.W. and them guys wasn't wanted here in the Springs."

Amid reports that strikebreakers were being recruited in San Francisco and Los Angeles, oil operators met in secret session at the Palace Hotel in San Francisco. As a result of the meeting the Oil Producers Association of California was formed. Rejecting any hint of government mediation, the association rallied under the slogan: "More business in government, less government in business."

In the weeks that followed, payments were started from the strike benefit fund: $10 a week to single men, $15 a week to married men. Businessmen of Taft subscribed $4,000 to the fund. Merchants of Maricopa issued a statement supporting the strikers: "All the oil workers are after is a square deal for all." Strikers requested a government probe of the strike, declaring that operators were painting them "worse than Russian soviets." The number of deputies in the oil fields increased to 1,075, and the strikers asked for troops, protesting many deputies were lease superintendents and foremen. Strikers paraded through the streets of Taft, 1,000 strong; a McKittrick oil worker hung himself. Friends said he was despondent over the course of the strike.

Warren H. Kraft, a chemical engineer who worked at Honolulu Oil's Taft office, recalled that E. R. Pratt, who was in charge of the company's gas compression plant in Taft, went to the main office one morning to make his report. The superintendent asked how his men at the plant stood on the strike. Pratt informed him in glowing terms of the men's loyalty to the company and assured him that they would stay on the job. Having finished their discussion, the two men drove by the plant. They found the place shut down and the employees out on strike.

Operators, professing to see mediation as the opening wedge in government control, rejected overtures from the union for a "settlement without victory." They warned that they would hold Kern County responsible for any violence or damage to their properties. They also filed more than 100 suits in Kern County Superior Court to force the eviction of strikers and their families from lease houses.

In the long run, the strike hinged on one question: what would happen to a well that was shut in? The state mineralo-

Coalinga camp, California Oilfields,
Ltd. 1911 photograph by Frank Foster.
(R. C. Baker Memorial Museum, Inc.)

gist had warned from San Francisco that wells might suffer damage. As the strike progressed, some wells were quietly returned to production by non-strikers; the wells had suffered no damage.

Early in November, two months after it had begun, the strike ended in victory for the oil operators. Wages were cut by one dollar a day, as the operators had intended, but working hours were not lengthened. Wells were returned to production against a background of complaints about the wholesale firing of men who had gone out on strike, an attempt to burn bridges on Highway 119 linking the West Side oil fields with Bakersfield, and sporadic shooting. Before the month ended, normal production had resumed.

If men looking for oil could have seen deeply into the earth, they could have saved themselves the time and expense of drilling dry holes. In the beginning they did not even have the tools or techniques to take full advantage of the window provided by the holes they drilled.

There was obvious interest in the layers of earth through which the hole passed. If an oil reservoir were not potent enough to flow when it was first penetrated, a sample taken from the hole might alert drillers that the drilling bit had found oil. They would be saved the loss—and the embarrassment—of walking away from a potentially productive well.

In time, men solved the problem of getting samples and making significant use of them. Other problems persisted. Sometimes they lost wells because they could not shut out water, often because they could not tell from where the water flowed. They could not necessarily tell if a well would be a producer short of the expense of running casing and attempting to complete the well. They needed information to tell them whether they were drilling a straight hole; when they learned how to find where the bottom actually was, they were only a step away from angling wells into pools that could not be reached from available drilling sites. They needed a method to tell them which sands contained oil and which contained salt water and some means to identify oil-bearing sands that might otherwise not be recognized. They also needed to find ways of correlating one well with another and of using the structural information thus developed to help find more oil.

Regarding samples, cable tool drillers claimed an advantage because they took samples each time they ran the bailer to clean out the hole. Such samples helped locate oil zones which did not flow but which could be shot with nitroglycerin to break up rocks and allow oil to flow.

Rotary drillers were handicapped by holes filled with mud. The mud prevented good oil zones from flowing, and zones were thus overlooked. Ditch samples were not enough to locate oil zones, so rotary drillers tried to take cores with a sharp, hollow-pointed tool which they dropped repeatedly on the bottom of the hole, jamming plugs of the formation into the hollow center. The method proved unsatisfactory, and they looked for other methods to bring samples from the bottom of the hole to the surface.

Shell Oil, whose headquarters were in The Hague, Holland, in 1919 and 1920 made extensive plans to attempt to take samples from shallow wells to be drilled in Southern California. The company borrowed a technique used in the coal fields of Holland, where miners employed hollow drills to cut samples, called "cores." They brought the cores to the surface in a "core barrel," the part of the drill used for sampling. In Holland, sediments were geologically old and compacted enough to be drilled with steel bits and to withstand disintegration from water circulation, so the coring devices proved effective. In Southern California, Shell contemplated taking cores in soft, unconsolidated formations at shallow depths. The company selected John E. (Brick) Elliott to head up the experimental coring operation.

To handle the coring assignment, Shell furnished Elliott the so-called "Holland" double-barrel core drills as well as compass-stratimeter devices to obtain the orientation of the core. This data helped to make use of information the core disclosed on the "dip," or angle, of the beds through which the core drill passed. The tools proved inadequate. The compass-stratimeter was too fragile for use with equipment employed in California drilling operations, and even after being redesigned, the tool would not stand up under heavy use and had to be discarded. Geologists devised a method, utilizing a transit and pipe clamps, of maintaining the orientation of the drill pipe while it was being extracted from the hole.

However, the coring equipment sent from The Hague had to be completely redesigned for operation under the mud fluid necessary in California drilling. The main roadblock, Elliott recalled, proved to be getting the geological department's suggestions accepted by the engineering department. Seldom was he able to get equipment designed as requested, suggested, and recommended. When efforts to obtain adequate

Fay Oliver, driller for Elliott Core Drilling Company, prepares to remove a core from the barrel after coring C.C.M.O.'s Well No. 96–8, 1927, (J. E. Elliott)

cores and information on orientation of the cores proved unsatisfactory, Shell abandoned the project. Elliott left the company, accepting a position as associate professor of petroleum technology at Stanford University.

By 1921, the majority of the wells in Southern California were being drilled by the rotary method, which was adequate for making hole but left operators without a way of locating the beds from which great amounts of salt water were infiltrating the oil sands. Elliott, despite the failure at Shell, continued to believe that the solution lay with the development of a double-cylinder core barrel which would permit an operator to distinguish between oil sands and water sands. Leaving Stanford after the spring semester of 1921, he attempted to sell his idea to oil companies, but met with a negative response. He gained encouragement from geologists and petroleum engineers, but both groups had yet to be "accepted" by oil producers responsible for oil field drilling practices.

Ward Blodget, a geologist-engineer with Chanslor-Canfield Midway Oil, urged Elliott to design and build a core drill. Blodget first advised, however, that Elliott attempt to get at least six companies to agree to take six or more cores over a test period of ninety days. The companies were to be charged $200 for each coring attempt to 2,000 feet or less, plus $10 per 100 feet in excess of 2,000 feet, with the stipulation that a core of at least six inches in length had to be obtained. In other words, no core—no money.

Elliott signed four contracts, one each with C.C.M.O., National Exploration Company (later absorbed by Shell), Union Oil, and California Petroleum (later part of Texaco). Another operator, E. J. Miley, simply made a verbal agreement; he later became one of Elliott's best clients.

The first attempt to obtain a core was made in the east end of the Huntington Beach oil field for National Exploration at Ashton No. 1 in August, 1921. With its success, Elliott Core Drilling got its start and incorporated the following year.

Up to this time few, if any, drilling wells were under continuous observation by geologists or engineers. Elliott offered the service of spending a week or ten days carefully observing the drilling of any well scheduled to be cored. Later he employed other geologists to help in the observation work. The added service stimulated interest in core taking and included contributing core containers, solvents for testing samples for shows, core tray tags, and written descriptions of the cores.

The most important part of Elliott's service was the supervision provided by technicians trained in taking cores. Elliott supplied his own core drillers. Core recovery, this crew learned, depended upon the rate of drilling, the amount and type of circulating fluid, and the weight on the core bit. Techniques had to be developed by trial and error. Core drillers had to be not only "top hands", but able to get along

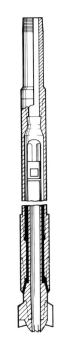

Elliott core drill.

Typical examples of the types of cores used for determining the orientation. (American Petroleum Institute)

with the operators' crews with whom they worked. Those chosen, Elliott recalled, were mostly "toolpusher" material, men capable of becoming drilling foremen. Some even had petroleum engineering degrees.

The Elliott crew of core drillers was responsible for many "firsts." They were paid toolpusher salaries for an eight-hour day plus a $10 bonus for each successful core. They were on call twenty-four hours a day, but they had little objection to this, for their average "take" ran about $1,000 a month, which was considered *good* pay. Elliott's first driller was Floyd Glass of Whittier, California, an intelligent and experienced man to whom Elliott credited much of the company's initial success.

Elliott confronted many unanticipated problems, not the least of which was the transportation of coring equipment to drilling wells. A used Ford truck was tried initially, but it lasted less than three months. Elliott experimented with Dodge trucks, which proved entirely too light for the amount of heavy equipment required. Finally, Floyd Glass, after fruitless experimentation with other trucks, began rebuilding used Cadillacs to serve as toolpusher trucks for carrying coring equipment. These became a huge success, being efficient and comparatively inexpensive. "Work for Elliott and drive a Cadillac" became a saying in the oil fields, and many companies eventually adopted the same practice.

No basic patents could be obtained on double-barrel core drills. These rights had been established in the operation of diamond core drilling in mining work and had expired many years before. Since there was no general patent protection available, Elliott recalled that his only means of keeping ahead of competition lay in superior service.

Early in the history of core drilling, some drilling contractors decided to refuse any coring attempts unless the oil company guaranteed the contractor against damage while coring, from accidents such as the loss of tools down the hole and the ensuing costly fishing job. Only the major companies and a few independent producers could possibly have made this guarantee. It appeared to be an impassable roadblock until Al Johnson, superintendent and part owner of Federal Drilling and an old friend of Elliott's, ordered out five core drills to five different wells, without guarantees, and told Elliott to inform independents that if they wanted cores, all they had to do was to contract with Federal Drilling. That solved the problem, and thereafter only Federal Drilling drilled Elliott wells.

The Elliott crew, during the early years, consisted of Floyd Glass, first driller and then superintendent; Bart Gillespie, Reid Granger, Pate Jordan, Sr., whose son later became track coach at Stanford; Jack Hartley, Larry Stiles, Ray McCoy, Fay Oliver, and Ed Spencer. Assisting during the early period were geologists Dwight Deckman, John McKenna, Harry Campbell, Al Carey, Bob Ames, and Tommy Cullen. Several became district representatives for Elliott Core Drilling in later years.

The record set by Elliott's team of drillers was an outstanding one. By 1922 the use of core barrels had become so widespread that they were accepted as a valuable tool in oil exploration and development. The net result of coring, of course, was that oil-finders now had cores to study. Early on,

Elliott Core Drilling took samples from this well, E. J. Miley's Athens No. 6, in 1927 when it was the deepest producing well in the world at a depth of 7,591 feet. (J. E. Elliott)

paleontologists recognized the value in the cores of foraminifera—microscopic, single-celled animals that develop a shell. Most forms are visible only through a microscope. When the animals die, their shells are left buried in the mud. There are many species, each with a characteristic shell; some species lived only in certain past geologic times. Thus, when certain shells, or assemblages of shells, were found, the geologic age

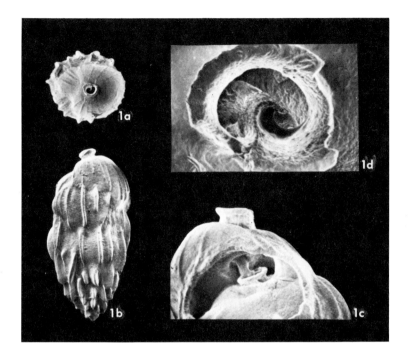

Side and top views of Uvigerina jacksonensis, *a variety of foraminifera important in petroleum exploration. These photographs were taken with a scanning electron microscope. (Jay Phillips, Museum of Paleontology, University of California, Berkeley)*

of the formation could be determined, and correlation became possible with formations not only at nearby drilling sites but also over great distances. Correlation of beds found in different wells became possible because of the foraminiferal shells present in the strata, and geologists had another tool to help them find the structures that might contain oil.

With the advent of coring, E. T. Dumble, who was head of Rio Bravo Oil, a subsidiary of the Southern Pacific Railroad, hired a paleontologist for the work of extracting and using "forams" for correlation purposes. Earl Gaylord, who was with Pacific Oil, another affiliate of Southern Pacific, reviewed the work that was being done and decided to start foram work in the San Francisco office of the company. He hired G. Dallas Hanna, who would later become head of the California Academy of Sciences. The work expanded so rapidly that Gaylord hired an assistant for Hanna, Herschel Driver. After Driver had been trained in the work, Clark Gester, chief geologist for Standard Oil, prevailed on Gaylord to let him have Driver, who then started similar work for Standard Oil. To their

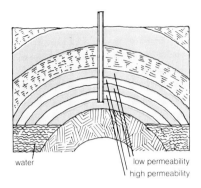

water | low permeability / high permeability

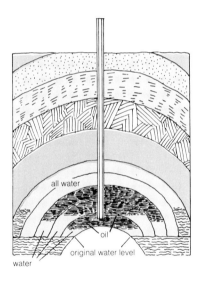

all water
oil
original water level
water

The diagrams show water conditions when a field is drilled (top), and the condition after "edge water" is encountered.

fellow oil men, the men who studied foraminifera through high-powered microscopes were soon known as the "bug men."

One problem that plagued oil operators, even with coring tools, was locating the point of water entry in a well. The stakes were high. If water could not be shut out, the well could be lost, and it was by no means easy to tell where water was coming from. A story that brought smiles concerned the field superintendent who told his superiors he was certain he knew where the water entering a particular well was coming from. He said it either came from around the shoe or out of the formation!

Two Standard Oil employees, R. D. (Pat) Elliott and E. J. Young, developed a solution to the problem. Their approach involved putting an electrolyte into the fluid in the hole. They then ran a probe down the well to measure electrical conductivity or resistance at various depths. When the probe indicated the electrolyte was being diluted, they learned the point at which water was entering the hole. They patented the approach, developing the "water witch" to serve as an additional tool for subsurface engineers. If water were coming from the bottom of the hole, engineers could plug it off. If it were coming from between zones, they could squeeze cement through perforated holes to shut it out.

From the earliest days of drilling, men sought a means of evaluating the formations encountered in a well without going to the expense of running casing. In the search for a method to simulate the conditions of a completed well, they recognized as early as 1867, when a patent was granted for a testing device, that any method of testing a formation would have to incorporate a packer—a device to fill the space in the well between the wall of the hole and the pipe—and a valve that could be opened to allow fluid from the zone being tested to enter the drill pipe so that the fluid might be recovered at the surface. The packer sealed off the interval being tested from the mud column above, reducing the pressure opposite the formation to atmospheric pressure to allow formation fluids to flow into the well bore.

In 1926, the Johnston brothers, E. C. and M. O., while

working in oil fields in the vicinity of El Dorado, Arkansas, conceived the idea of using a spring-controlled valve with some sort of packer below to test formations that because of their irregular nature and thinness were costly to test with casing. The approach involved drilling a reduced hole of smaller diameter into the zone to be tested, leaving a shoulder above the zone. They devised a packer, conical in form, to be run on the end of the drill pipe, and ran the spring-controlled valve above the packer. Material for the packer was cut from discarded belting—in plentiful supply in oil fields—chosen because it would rupture more easily when the tool was pulled from the hole. Other parts of the tool included a heavy spring from a railroad boxcar and a poppet valve that rose perpendicularly to and from its seat. To prevent the valve from opening while the string of pipe was being run in the hole, the Johnston brothers fastened metal straps to the outside of the housing so that the spring could not compress. The closed valve kept drilling fluid from entering the pipe. In field runs, the testing equipment worked.

There were limitations, however. The set-up did not permit much spudding—up and down yo-yo action—to set the packer on the seat or to pass crooked sections of the hole. To get away from use of a valve activated by the movement of the pipe, and hence subject to open while pipe was being run in the hole, the Johnstons added to the assembly a trip valve activated by dropping an iron bar on the plunger of the valve.

Equalizing pressure above and below the packer after a test created a problem. Pressure below came from the formation fluid, while that above was the weight of the whole column of drilling fluid, often many times greater than the formation pressure. There were instances in which five-inch drill pipe collapsed under the strain of trying to pull the packer loose. As the depth of testing increased, the problem became more acute. In 1931, the Johnstons put into use an equalizing valve which allowed pressures above and below the packer to be equalized after the fluid sample had been obtained.

California state law in the interest of oil conservation required operators to test water shut-off after they had set casing, with the test to be witnessed by a representative of the

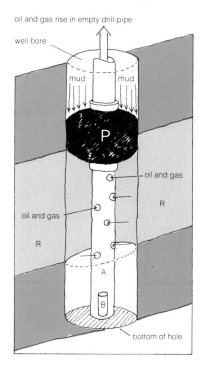

Diagram illustrates main elements of a testing tool. Expansion of rubber packer (P), due to weight of drill pipe, separates heavy mud column from reservoir fluid/pressure system. Perforated anchor (A) below packer permits reservoir (R) fluids into empty drill pipe. They rise toward the surface. Pressure gauge or bomb (B) registers flow and shut-in pressures when valves open and close tool.

oil and gas rise in empty drill-pipe

well bore

mud   mud

P

oil and gas

R

oil and gas

R

A

B

bottom of hole

Babe Allison, Doc Knowlton, and Brick Elliott inspecting last core of Miley's Athens No. 6 in 1927. (J. E. Elliott)

state's Division of Oil and Gas. The water shut-off had to be approved before additional work could be done on the well. The test was to determine if water above the oil sand had been effectively excluded from the well. If the water were not excluded, it could enter the oil sand and cause some of the oil in the reservoir to be unrecoverable.

Prior to 1931, operators after cementing casing made water shut-off tests by drilling out cement not more than five feet below the casing shoe, then bailing or swabbing the fluid down to a required depth and allowing the well to stand twelve hours. Afterwards, the crew ran in with a bailer to locate the level of the fluid. If the rise amounted to a rate of more than one and a half barrels in twenty-four hours, the test had to be repeated. The operation was time-consuming and costly.

It occurred to more than one operator that if he had a packer small enough to operate inside casing, he could make a test, similar to a formation test, to determine much more quickly if water were entering the hole. M. O. Johnston built such a packer in 1931 and successfully tested it. The state accepted the new procedure, which only required from three to eight hours and represented a substantial saving in rig time.

Sometimes drilling bits developed a tendency to wander,

and the resulting crooked hole posed problems. If there were a dog-leg in the hole, the drill pipe might cut over to straighten the hole and the resulting key seat, or the place with pronounced vertical channeling of the walls, formed a hazard; the driller might try to pull the string of pipe from the hole, pull the drill bit into the key seat, and find his string of pipe stuck. If a hole were too crooked, it might break the string of casing run to complete the well.

Before the invention of survey tools, the deviation of wells from vertical, whether drilled by cable or rotary tools, had long been known from such indication as line-cut casing, the behavior of both rigid and flexible bailers, key seats, and other phenomena. It was important to operators to know the amount of deviation from vertical, as well as its direction. Getting such information called for the development of survey tools that could be run on wire line or drill pipe.

Surveying the amount of deviation was the easier problem to solve. Miners for many years had used acid bottles in which hydrofluoric acid etched a line on a glass container to measure inclination from the vertical. The first step toward adapting the technique to the oil fields appears to have been made by several people in Southern California in the middle 1920's. Commonly, an acid bottle was run on a wire line or dropped as a "go-devil," that is, carried with fluid down the drill pipe. Zeb Dyer designed and ran one of the earliest acid bottle surveys at Signal Hill about 1924. Many variants in carrier design, spring systems and "readers" were used.

To provide a record of direction, men used two main approaches. One involved a system of reflecting mirrors, a light source and a film on which the record was made. Alexander Anderson, who had had some experience surveying diamond-cored holes in The Rand, South Africa, integrated the devices into a surveying tool after he came to California early in the 1920's. He had the idea of surveying oil wells using an oriented drill pipe method. He filed his first patent application in August, 1924. Several patents were issued to him in the period from 1930 to 1933. Sperry Sun developed its own method using a gyroscopic compass to determine direction, with a photographic method to record it.

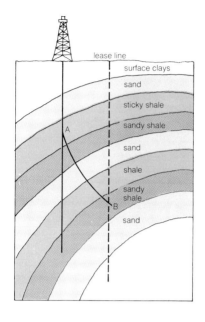

The whipstock (right,) directed the drill bit at the desired angle for directional drilling (diagram above,).

With instruments like Anderson's and Sperry's, it was possible to determine the location of the bottom of a well. The next step involved directional drilling, which put the bottom of the hole at a selected site away from the place the hole began. Sometimes a well could not be economically or physically located over the targeted structure, making it desirable to develop directional drilling capability. To drill directional holes, men turned to the whipstock, a beveled bar of steel placed in the bottom of the hole at the point where deviation from a straight hole was desired. The whipstock directed the drill bit at the desired angle. The tool took its name from the long taper at its upper end which resembled a buggy whipstock. The first directional drilling was done by Eastman Oil Well Survey about 1930, using a McVickers removable whipstock. Superior Oil discovered the Huntington Beach offshore pool in May, 1930, by directionally drilling the Babbitt No. 1 from an onshore site.

Of all the advances being made at about this time, perhaps none would have a greater impact than the electric log, a simple tool that opened a whole new field to geologists and engineers. Electrical effects in mines had been noted as early as 1830, but the development of electric logging started in 1912 when Conrad Schlumberger, of France, began systematic studies of electrical resistivities along the surface of the earth. On September 5, 1927, Conrad and his brother, Marcel, recorded the first experimental electric log in a bore hole at Pechelbron in the Alsace district of France. The Schlumberger brothers lowered an instrument in the hole to measure electrical resistivity of strata at fixed points and later plotted the resistivities as a graph against depth. The event was probably the single most important event in well-logging history. Afterward, the Schlumbergers devoted their entire time to electric logging, bringing the method to Venezuela in 1929 and, in the same year, to the United States.

The method of operation consisted of lowering a special electrical instrument into the hole on a wire line to obtain electrical resistivity measurements of the strata in the hole. Readings were recorded on a long film with depth markings corresponding to depths at which readings were made. Since

189

each bed had its own peculiar profile on the electric logs, logs enabled geologists to distinguish between sands and shales and to correlate beds from one well to another. This gave them valuable information with which to determine the subsurface structural features of the area. The electric log provided engineers with a tool by which they could distinguish sands containing oil from those containing salt water. It gave them information that helped tell how deep to drill, where to set casing, and where to stop before getting into bottom water.

Jacques Gallois made the first electric log run in California at Shell's Boston Land Company No. 1 at Westhaven near Huron, Kings County, on August 15, 1929. A few years later, Schlumberger opened an office in Long Beach with Roger Henquet as supervisor and one in Bakersfield directed by

Gilbert Deschatre and Jacques Gallois with the first electrical logging truck in California, 1933. (Petroleum Club of Bakersfield)

Gallois. Electric logging became a science, and the electric log became as important a tool to the oil operator as the rotary rig.

In summary, the advances of the 1920's and 1930's meant that when an operator started to look for oil, he could better tell where the drill bit actually was in relation to where it was supposed to be, what the bit was passing through, and whether he had found oil or gas in commercial quantities. With new tools and techniques, there was less chance he would walk away from a potentially productive sand, and there was the added hope that, though a well might be dry, the geologic information secured from it might be used later to point the way to another drill site where he might bring in a producing well.

13 Natural Gas: From Nuisance to Asset

Gas flares burning in the booming Signal Hill field often provided enough light to permit men to continue working at night without any other source of illumination. In the early days of the oil industry the presence of natural gas in producing areas appeared as more of a nuisance than an asset, and those who produced it worried more about getting rid of it than about using it. The simplest method for gas disposal proved to be piping it to a riser and burning it off into the atmosphere.

The presence of gas threatened to cause explosions, however, and Garth Young, who first went to work for Signal Gasoline in 1922, recalled an especially spectacular one. "Signal Hill was alive with drillers, pipeliners and wooden rig builders. Rooms were scarce so I holed up with a few other roughnecks in the garage on Cherry Avenue on the south slope of the hill. We all boarded in the same place, which served meals at all hours. Not long after my rent started, Union's well just behind the garage hit a big gas pocket. Pipe blew out of the hole and wrapped around everything nearby including my garage quarters. No one was hurt, but mud and water from the well fell on the roof, leaked in, and almost collapsed the structure. The well blew wild for several days, but fortunately did not catch fire. However, enough mud cascaded down the hill to cover the Pacific Electric train tracks four feet deep, suspending train service for a week or two. As suddenly as it blew out, the well sanded-up again and flow stopped. Then we all moved back into the garage and resumed sleeping there a few hours each night," he said.

Despite its hazardous properties, natural gas became increasingly recognized as a potentially significant energy source. Long Beach, the area in which the Signal Hill field was located, had supplied gas to its citizens since the turn of the century. At first, this product was gas manufactured from coal, and then a half-and-half mixture of manufactured and natural gas. Finally, in 1915, straight natural gas was brought into the area from the Fullerton oil field. The discovery of the Signal

E. P. Reynolds and engineer Ronald W. Heath (right) with Signal's gas machinery at Signal Hill, 1924. (Burmah Oil and Gas Company)

Hill field, of course, provided a closer source of natural gas, and several operators—especially Samuel Mosher and Robert Bering—searched for ways of making profitable use of the abundant resource.

As produced at the well, natural gas may contain some water which is corrosive and may form compounds that freeze and constrict gas lines. The water should be removed at the earliest possible moment, and this was done at Signal Hill by exposing the stream of gas in an absorption tower to glycol or other substances which absorbed the water without entering into any chemical reaction with the gas. The gas bubbled up through a counter-stream of glycol or something similar, and when it had lost all or most of its moisture, it was termed "dry" and ready for transportation by pipeline to a processing plant. The earliest absorbers, R. S. (Bob) Tulin of Shell Oil recalled, "were packed with bailing wire, wooden grids, expanded metal, and almost anything coarse enough to allow gas and oil to pass through them."

Oliver C. Field, who later founded O.C. Field Gas, recalled the origins of the early refining process. "The utility companies demanded dry gas of uniform quality," and this demand forced the operators to develop "efficient methods for removing the condensable gases and refining them. Chilling and scrubbing," he said, "were the original processes applied to this purpose, scrubbing being a system of holding the gas more or less in suspension at low temperature so that liquefiable fractions and other undesirable refuse would be dropped out." They discovered, Field said, that the fluid "recovered from the treated gas was a very volatile, or 'high test,' gasoline which could be blended into straight-run gasoline at the refinery to give it better starting quality and better all around volatility."

Field explained that "the compression method had one disadvantage. It was not sufficiently selective and tended to condense wild, unstable 'ends' along with the more desirable fluid hydrocarbons. Besides producing a mixed commodity, parts of which had unwanted qualities, the method of separation was also expensive. The installations were costly and this, together with a lack of flexibility in the matter of

Gasoline truck taking on casinghead gasoline 1923 from Signal's Plant No. 1 at Signal Hill. (Burmah Oil and Gas Company)

selection, made it imperative to find a better method. A better way of condensing and separating the usable fractions of the relatively richer gas from deeper wells was found to be the process of absorption, involving mixing the gas from the well with a neutral absorption oil which might subsequently be distilled to recover the natural gasoline. The job was accomplished in absorption towers, which were nothing more than tall vessels with layer upon layer of trays inside."

The early absorption plants were primitive. "During one period of construction," Garth Young recalled, "it was decided by Signal to cover the still shell with magnesia blocks for insulation. The blocks were neatly stacked up behind the rear head of the still so that work could start next morning. In the late afternoon, pressure built up in the still and before it could be relieved by hand blow-down, the head ruptured, blowing bits of magnesia around and about until it looked like a snow storm. The still head had left the shell, cut off two of the four dry gas lines on the absorbers, cartwheeled through the back fence landing in the street one block west of Orizaba." Young said that "luckily, there was no fire, so no one was injured, but a few days' production was lost."

There seemed never to be a dull moment. Garth Young recalled that "fires and blowouts became more numerous, production increased at a rapid rate, natural gas was blowing out of separators faster than gas lines could be laid to get it into the plant. Signal had installed two Worthington belt-driven gas compressors to pick up low pressure gas but with added gas from nearby wells the pressure kept increasing to such an extent that the suction pressure on the intake header was often higher than the discharge pressure.

"A number of the new wells were drilled by George F. Getty, the first, that is," Young said, "He was J. Paul Getty's father. This old pioneer was a typical oil man with leather puttees, jacket, cap, and steel-blue eyes. He told Signal to hook up to his wells and take the gas. There was no contract. Just pay him the going rate for such gasoline as we sold. There was no such thing as dry gas sales in those days. Mr. Getty covered a lot of ground, day and night. His wells were the most pro-ductive ones we tied into, and he would drive into the plant yard frequently, looking for me and his gas charts. He didn't know how to read a chart for gas volume, but he did know how to read the pressure recording line showing whether the well was steady or surging. Each chart showed continuous recorded pressure around the clock. To him this was a very significant clue to the behavior of his wells. We became good friends over the few months, during which time he would hardly wait for a gas chart to be installed on every new well so that he could watch the well behavior for himself," Young remembered.

Garth Young also fondly recalled the "Spud-In Cafe" on Obispo Street, run by "Nick the Greek." "The cafe was a tent stretched over a wood frame with counter and kitchen all in one," Young said.

"It ran twenty-four hours a day. Nick never seemed to sleep. He was always ready with short orders or anything else that was wanted. Coconut cream pie was his specialty. About twice a week, sometimes in the daytime, sometimes at night, the place was robbed. The robbery was so common that no one paid much attention as to the method of procedure. Nick, however, got tired of the raids after awhile and decided to do something about them. He made a deal with the city of Signal Hill police to put in a red light on a pole a block from the cafe, with a switch in his place so that he could alert the police when a robber had just left his place with the receipts. They caught every one of the robbers within a few weeks. That ended the problem, and Nick thereafter enjoyed a life of peace and security," Young recalled.

Historic "look-box" enabled prospective buyer of Signal stock to see actual throughput of gasoline plant. (Burmah Oil and Gas Company)

Before the boom at Signal Hill, a few utility companies still used some manufactured gas. The cost of the manufactured commodity was much greater than that of natural gas, so a market quickly developed for the latter. Previously, the only outlet for dry gas had generally been its use as oil field fuel, and any loss or unreported usage was not considered very important. With growing demand for dry gas and natural gasoline bringing an attractive price, the situation changed. All sorts of deals and arrangements were made between oil producers and gasoline manufacturers. A royalty of thirty-five percent soon leveled off, but varying types of bonuses were given to producers with accessible well locations along gas-gathering lines already laid. The bonuses varied from free fuel gas for drilling to surface pipe, and in some instances, even cash.

Some individuals saw no harm in helping themselves, without asking, to the gas that flowed through field gathering lines. Ron Heath, who worked in Signal's original plant, recalled that "the usual procedure for checking the unauthorized usage of the gas from the field lines was first to notify those who were really entitled to it that at a certain time the line would be closed off for a test. It then became a simple matter to determine whether or not unauthorized persons or concerns were tied into the line. When they were discovered, they either signed a contract or the illegitimate connection was detached or closed off. There was one line Signal shut down and found that it was being tapped by a restaurant and two rooming houses. Usually in such cases the guilty parties claimed to have been given permits by some persons whose name they couldn't remember, but certainly with more authority than those now challenging them. Needless to say we rarely lost the argument, but he lost his assumed privilege and the gas, too, if he didn't sign in the space provided," Heath recounted.

"While it was common to find people helping themselves to our gas, there was at least one sort of reverse situation," Heath said. "A well in Cerritos had come in and produced about 3,000 barrels of oil. The first shipment from the well had been inadvertently run into our dry gas line. We had approx-

(Burmah Oil and Gas Company)

imately 100 boilers being fed by this dry line. When the oil hit the boilers, we had the first 'smoke in' in oil history. We knew at once what had happened, but several hours elapsed and a lot more smoke was produced before the source of the oil could be pinpointed."

If some helped themselves to gas, others helped themselves to automobile fuel. Natural gas is an elastic substance of which a large volume can be compressed into a very small space and when it is compressed at reasonably low temperature, the heavier components are liquefied. Because of this, lines carrying gas from the separator traps in the field to the gas plants were equipped with drip boxes, or pots, in which the condensate might gather.

The condensate had various uses. Carlton L. (Cop) Case recalled that, while working on the White Star lease in the Little Sespe Creek area near Fillmore, he gathered drips from gas fuel lines during the winter for use in starting the Fairbanks engine powering the jacklines on the lease.

"After pouring some of the volatile drips in the power cylinder," he recalled, "we turned it over several times by hand and set it in firing position on the compression stroke. We then ignited it with a 22-blank cartridge or the head of a big kitchen match in the firing pin device that protruded through the cylinder head. It was almost like serving in the artillery."

The condensate was found to be excellent blending material for aviation or automotive gasoline. In 1919, after the war, Case worked in Ventura Oil's gasoline plant on the Shiells lease at Fillmore. "The liquefied petroleum gas from the accumulator," Case recalled, "was blended with the water white distillate from a steam stove operating on the lease to make a fluid of approximately 72°-gravity. This was run through fuller's earth filters to improve color and remove moisture. The filters were made from twelve-inch pipe, and the fuller's earth was rejuvenated by the application of heat. The only trouble with the process was that the product had an end point of about 550° Fahrenheit due to the presence of heavy distillate. This blend, however, sold from ten to twenty cents a gallon. We furnished the fuel for most of the race cars at twenty cents a gallon in drums, after the old board track in Beverly Hills was started

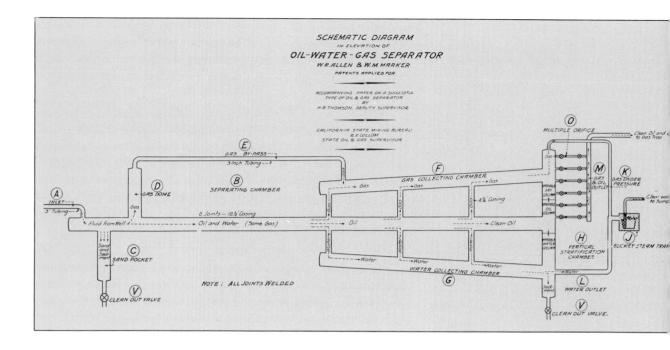

Schematic diagram in elevation of oil-water-gas separator used in refining natural gas.

up. The high-speed Dusenberg engines were just becoming popular in the United States, and drivers liked this fuel."

It did not take long for some field workers to recognize the condensate as a convenient source of energy. In some areas it was at first permissible to use "drip gas" for company cars, but when the practice was determined to be hazardous, it was officially stopped. Others, however, continued to fill their tanks with drip gas. Since the theft was performed in the small hours, the condensate acquired the name, "midnight ethyl."

Not all of the "products" that came from Signal Hill's absorption plants worked as well as the drip gas that never was inside any plant. Marion Arnold, who worked in Signal Gasoline's plant and later became manager of gas operations for Richfield Oil, recalled one special commodity; it consisted largely of carried-over absorption oil and heavy gasoline fractions that were accumulated by the plant and sold to an itinerant merchant who in turn sold at least some of it to orchardists for firing smudge pots during winter cold spells. Once, a new operator, unfamiliar with the transaction, had in error furnished the merchant with a load of badly contaminated, low-grade motor oil that would not even burn. The merchant then sold it to a customer who because of it lost his entire citrus crop.

The natural gas and gasoline industry encountered many problems. One involved an insufficient understanding of the apparent volumetric shrinkages that took place during transportation. Paul Barton, with General Petroleum in 1920, recalled what happened at his company's compression plant in the Midway-Sunset field near Taft. "The gasoline was produced in a battery of vertical tanks, equipped with gauge glasses," Barton explained. "It was measured under pressure, and then pumped to the terminal where it was similarly gauged and measured. A loss of considerable magnitude was consistently observed, and a friend of mine spent three months walking the line trying to find the source of the loss. Finally, the company decided that there was something inherent in the nature of the wild gasoline which accounted for the loss.

"Some years later," Barton said, "I was talking to an old operator who had been at the plant at the time, and he told me confidentially that there were cones of solid ice in those vertical tanks that were sold over and over again to the gasoline purchaser." The ice cones were formed by water carried over and frozen during sub-cooling by evaporation of the gasoline.

George Tyler, who served for many years as secretary of the California Natural Gasoline Association, recalled how Pacific Gasoline ran natural gasoline into Associated Oil's crude tanks at Fellows, preparatory to sending it by pipeline to Associated's refinery in the San Francisco Bay area. Everyone was disturbed because "before" and "after" gauges always showed a serious loss, even though measured at, or calculated to, the same temperature and pressure conditions. Not until years later did researchers discover that the blend of such commodities does not yield a straight arithmetical volume increase. In other words, if one gallon of natural gasoline is mixed with one gallon of crude oil, they blend in a way that does not yield two gallons of fluid.

One of the main problems that faced the natural gas industry concerned the types of absorbers used in the refining process. Absorption towers of infinite variety had been built, using all sorts of internal structure and contrivances to accomplish the mixture and ultimate separation of oil and gas. These towers were followed by various types of bubble tray columns which were put on the market by a number of designers and builders, each making certain claims of performance but offering little reliable information in support of the claims.

Bob Tulin recalled that "early in 1925 an investigation was begun at the Signal Gasoline Company's plant, located on the Denni lease at Signal Hill, for the purpose of determining the efficiency of various types of available absorption towers. Arrangements were made with the manufacturers of the newer absorbers then in use, and four makes were installed in parallel at the Signal Hill plant." The absorbers included a surface content absorber made by C. F. Braun, a perforated plate oil-froth tower made by J. A. Campbell, a foam absorber made by Newton Process Manufacturing, and a bubble cap absorber made by Southwestern Engineering.

Tulin explained that the testers hoped that "by varying gas rates, oil temperatures, absorber pressures, and by taking a series of samples from each of the absorbers, in accordance with the variations, it would be possible to determine once and for all which type of absorber was most efficient.

"In order that the tests might be entirely impartial and more comprehensive as well, an invitation was extended by Signal Gasoline to those who were interested in the problem, to participate in all or any of the tests," Tulin recounted. "After several months of testing and correlating the results by the various individuals, it was found that although the differences in performances between the absorbers were slight, the real difference lay in the lack of uniformity in the results obtainable by the test methods. It was thus decided to defer further absorber tests until the discrepancies and differences in testing methods could be reconciled and reduced at least below the variations in absorber efficiencies. So it was that the cooperative effort to solve a common problem turned attention to the test methods that were in use and the need to devise new and better tests."

Making fuel for airplanes captured the imagination of refiners. When the "Angeleno" set a new endurance record in July, 1929, Richfield's "aerial service station" refueled the Angeleno in midair thirty-one times. (Atlantic Richfield Company)

Among those who engaged in the project were Paul Barton, general superintendent of Signal Gasoline; Harold Linhoff, in charge of that company's plant; and Les Warner and Marion Arnold in the plant laboratory. Others who participated were Julian Campbell and Henry Wade of Lomita Gasoline; Roy Huff and A. J. Hutchinson of Pacific Gas; Dan Wolfe and Walter Dayhuff of Standard; Stewart Watson and Ed Cummings of Shell; Andy Kirk and J. L. Patroso of Export Refining; Ray Wheeler of General Petroleum; M. H. Scott of Carbide & Carbon Chemical; and D. H. Cushman of California Petroleum.

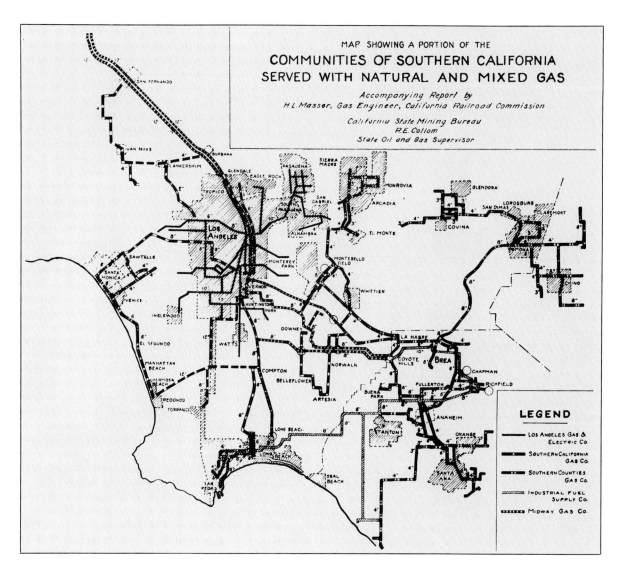

MAP SHOWING A PORTION OF THE
COMMUNITIES OF SOUTHERN CALIFORNIA SERVED WITH NATURAL AND MIXED GAS

Accompanying Report by
H.L. Masser, Gas Engineer, California Railroad Commission

California State Mining Bureau
R.E. Collom
State Oil and Gas Supervisor

LEGEND

————— Los Angeles Gas & Electric Co.

▬▬▬▬ Southern California Gas Co.

▬▬▬▬ Southern Counties Gas Co.

░░░░░ Industrial Fuel Supply Co.

▪▪▪▪▪ Midway Gas Co.

Some of those involved in the project had a short time before decided to meet regularly for informal discussions of operating and testing problems among the neighboring technical personnel at Signal Hill. Shell had a field headquarters on Willow Street, and Stewart Watson made arrangements for the group to meet in the company's social hall. Among those attending the early meetings were Marion Arnold, Harold

State Oil and Gas Supervisor's map of natural and mixed gas transmission lines, Los Angeles vicinity, 1922–23.

Linhoff, Julian Campbell, Oliver Field, Stewart Watson, Clare Gard and later, as the idea of establishing a formal technical association developed, Reid Gorman, I. B. Funk, R. E. Beckley, George Ratcliff, Walt Dayhuff and Herb Eggleston. From these get-togethers and the project to study absorbers came the formation of the California Natural Gasoline Association, which served as a technical forum and agency for promoting the adoption of standards for reference data, testing procedures, measurements, design, and related matters.

Elected as the first president, Paul Barton recalled that "in November, 1925, the American Petroleum Institute held its annual meeting in Los Angeles. As president of the newly organized C.N.G.A. I promoted a formal fall meeting of the association, and we prepared to hold our first full-blown technical session to coincide with the American Petroleum Institute meeting. Its success was unqualified and our association was fully launched."

14 Going to Sea for Oil

The Southern Pacific liked to call its coast line the "Road of a Thousand Wonders," and at Summerland, near Santa Barbara, travelers could see derricks on wharves over the sea, pumping oil from beneath the ocean's floor. The search for oil had led wildcatters into the surf and beyond.

At Summerland, H. L. Williams drilled the first well on-shore in 1886, turning up shows of light oil in the 455-foot hole before the casing parted and the well was lost. He drilled and completed a second well, and others joined the onshore play, bringing in wells from depths of eighty to 150 feet.

Natural gas sand existed in this area at about 100 feet. In spite of the shallow depth, the gas was under enough pressure to throw mud and dirt forty feet into the air when drillers tapped the sand. An operator named Cone drilled three wells of two and one-half inches in diameter into the gas sand and supplied twenty families with fuel. The Darling brothers drilled two wells into the sand, the first in 1891, and produced enough gas to supply seventeen families. Pressure in the Darling brothers' wells was eight pounds per square inch.

As development of the Summerland field continued in the 1890's, it became apparent that oil sands dipped beneath the ocean, and operators quickly followed them from piers extended into the water. The field was a townlot development with at one time nearly 100 operators represented in the play. In short order, there were fourteen north-south piers extending into the ocean, with three connecting wharves. The Southern Pacific Railroad built the longest of the piers, 1,230-feet, to get in on the drilling boom.

Summerland offshore drilling, c.1890. (C. C. Pierce Collection, Title Insurance and Trust Co.)

Wells were started from the piers through a section of large casing—conductor pipe—driven into the sandy ocean bottom to shut out sea water. Inside this pipe cable tools drilled the well, with the rig being set on temporary planking on the pier. Completed wells were pumped with jacklines, and produced an average of between one and two barrels a day of approximately 14°-gravity oil that brought as high as eighty cents a barrel. The oil was black oil similar to that produced in the San Joaquin Valley's Kern River field.

Piers proved difficult to maintain through stormy periods, and the collapses caused by sea action broke off casing below the water level, admitting great quantities of ocean water to oil sands. Production of the wells was too small to justify the cost of drilling them from piers and thus prohibited further extension into deeper waters. The drilling of wharf wells ceased about 1899. In most cases it is doubtful that the original cost of piers and wells was ever repaid.

Offshore drilling took place from piers built out into the ocean. Summerland, 1890's. (Huntington Library)

Nevertheless, where piers survived, wells continued to produce, and California Liquid Asphalt operated a refinery in the field until 1910, using a considerable portion of the oil for the manufacture of asphalt, with the balance going to Santa Barbara by rail for fuel. In 1903, 198 wells were on production in the Summerland field; an additional 114 were idle, and some 100 had been abandoned. Producing wells averaged 1.82 barrels a day per well. After California Liquid Asphalt's refinery ceased operation, B. S. Bennett constructed a plant and for a number of years continued in operation, making asphalt and heavy distillate from about half the field's output, with the remainder of Summerland's production finding a market in Santa Barbara as fuel and in the adjacent district as road oil.

Given the early example of Summerland, it was inevitable that wildcatters sooner or later would be drawn back to the ocean.

In 1921, the state of California passed a tidelands leasing act authorizing the Surveyor General to issue prospecting permits on state-owned tide and submerged land. Under terms of the act, the state gave the permittee the right to a lease provided the permittee made a commercial oil discovery.

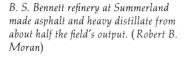

B. S. Bennett refinery at Summerland made asphalt and heavy distillate from about half the field's output. (Robert B. Moran)

Elwood field in Santa Barbara County proved to be another pier development in the 1920's. (Division of Oil and Gas, Long Beach)

Wildcatters sought and received permits to drill for oil off the counties of Orange, Ventura, and Santa Barbara. They used the same tools and methods employed onshore, simply building piers to serve as a base for drilling operations. The public responded with concern and resentment over this use of the coastline. Between 1926 and 1928, the Surveyor General deferred issuance of several hundred tideland prospecting permits. Applicants went to court. When the State Supreme Court on December 31, 1928, rendered a decision requiring the Surveyor General to issue permits to qualified applicants, the state's legislature countered with a moratorium, prohibiting the Surveyor General from granting any lease or prospecting permits. The legislature followed up by enacting a law prohibiting the issuance of leases for the production or extraction of any mineral from tide or submerged lands.

Meanwhile in the Elwood area, twelve miles up the coast from Santa Barbara, Rio Grande Oil developed a prospect and acquired onshore leases from private owners. Barnsdall Oil had an idle drilling rig in the area and agreed to drill a wildcat on Rio Grande's Luton-Bell lease, paying half the cost of the well in return for half interest in the various leases held by Rio Grande. In July, 1928, Barnsdall, as operator, brought in the Luton-Bell No. 1 at an onshore site on the bluff overlooking the ocean. The 3,208-foot well flowed at a rate of 4,300 barrels a day from the Vaqueros sand, proving up a major oil discovery.

Under then existing state law, upland owners of beach frontage had preferential rights to tidelands mineral claims for a period of twelve months. If no claim were filed by the owners during that time, anyone else might do so. Rio Grande duly filed on the tidelands fronting the Luton-Bell lease, completing on the filing form the year, month, and day of filing, but failing to register the time of day, as required by law. The tiny omission, if not discovered in time, was enough to invalidate a tidelands mineral claim.

While Rio Grande and Barnsdall continued to bring in big wells onshore, a Los Angeles attorney named David Faries spotted the flaw in Rio Grande's filing and silently watched the months slip by without corrective action on Rio Grande's part. Exactly one minute after the twelve–month period expired, Faries and associates filed on offshore leases at Elwood and took legal control of the properties immediately adjacent to the growing number of fast-flowing wells onshore. Faries, in turn, leased a portion of the offshore acreage to Honolulu Oil.

"Rene Broomfield, general manager for Barnsdall, looked upon this as an unfriendly act," H. M. Van Clief of Honolulu Oil recalled. "He refused to grant Honolulu any type right-

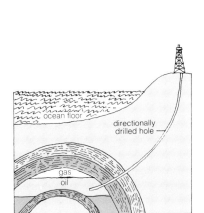

Diagram indicates directional drilling from onshore site to offshore oil.

of-way across Barnsdall's properties, which extended for a mile or more on each side of the Honolulu tideland permit."

The action posed problems for Honolulu. Legal counsel advised that condemnation procedure would take longer than the time specified in the permit drilling clause. Honolulu resourcefully borrowed a ship from its sister organization, Matson Navigation, and anchored the vessel offshore for use as a base of operation.

"A raft was outfitted with pile-driving equipment," Van Clief recalled, "and several attempts were made to drive piling

First oil wells in the ocean were drilled in the Summerland field, where C. C. Pierce took this photograph in 1902. (Title Insurance and Trust Co.)

At Huntington Beach oil men drove trucks up to the rig, but their access was limited to low tide. (Fritz Ripley, Petroleum Production Pioneers Collection, Long Beach Public Library)

in the calm water behind the breaker line. All efforts ended in failure. Meanwhile, at Honolulu's request, the Surveyor General had established a mean high tide level across the Barnsdall property fronting the ocean. The beach below this line became the only access by land and of course was usable only at low tide as the high tide beat against the cliff-rimmed shore. So the pile-driving raft was beached and stabilized, and a pier was started from there. Drilling equipment was dragged in by tractors, and in due time drilling operations began.

"The drilling crews changed tour at low tide instead of at

the usual midnight, 8 a.m. and 4 p.m. intervals. All suppliers and services had to follow the same schedule because Barnsdall had armed guards posted around the clock on their side of the mean high tide line to make sure there was no encroachment by Honolulu personnel or equipment. When it became apparent that the 'no right-of-way' policy was ineffective, Barnsdall allowed Honolulu a restricted use of their roads and finally abandoned all restrictions," Van Clief said. The Elwood episode created no feud, and the two companies later took part in a joint venture in Canada.

During the tidelands drilling operation, the Matson ship continued to serve as an operational base. Three-pronged animosity developed among men working on the project. The Honolulu drilling crews and the marine construction contractor's crews looked upon each other as underworked, overpaid, bungling braggarts. Both believed the ship's crew existed to cater to their needs. In turn, the ship's crew looked upon the oilmen as an inferior group of landlubbers unworthy of a berth on a sea-going vessel. It took a bit of diplomatic finesse to avoid an open rupture, especially when all efforts at sea-going pile driving had failed. The feeling permeated higher levels.

"When the Honolulu manager advised the ship's captain that he was free to go home," Van Clief recalled, "the captain upped anchor and took off for San Francisco. Fortunately for Matson, the ship was detected before it entered the harbor and was sent back to Santa Barbara for proper clearance. The Honolulu manager, R. R. (Bob) Maguire, knew nothing of maritime law though he found out something about it the next day after releasing the ship. Mr. Diericx, Honolulu president, and an ex-Matson executive, called him on the telephone and really chewed him out for sending the ship to San Francisco without first clearing the nearest port of Santa Barbara.

"Bob proclaimed his ignorance of such procedure but Mr. Diericx's only reply was, 'Any damn fool knows better than to do such a stupid thing.' Mr. Maguire was at a loss to understand why he should be blamed. During the whole operation there had been many costly errors, but no reprimands or complaints from top management until this incident involving

Swimmers and sunbathers share the coastline with oil wells, Huntington Beach, 1928. (City of Huntington Beach)

the ship's return. Despite the manager's ignorance, there was no mystery to even the youngest roughnecks in the drilling crews. This was the captain's way of getting even for past insults to his ship and crew," Van Clief explained.

By the end of the 1920's operators had chased oil quite a distance from the shoreline of Southern California. Drilling equipment reached out from the shore more than a mile at Huntington Beach and other locales. Oil sands undoubtedly went further out, these operators thought, but at that time, their equipment did not permit them to go farther out to sea. Not until after World War II did the industry develop the floating rig that enabled wildcatters to break their link with the land and make offshore drilling the enterprise we know today.

15 The Deepest Oil Well in the World

The first clue that guided men in the search for oil was the oil seep. Edwin L. Drake in 1859 drilled on an oil seep on Oil Creek in western Pennsylvania, bringing in a thirty-barrels-per-day well that gave birth to the oil industry.

Six years later, a Canadian geologist, T. Sterry Hunt, of the Canadian Geological Survey, concluded that four conditions were necessary for oil accumulation: a source bed containing material from which oil could be formed; a bed of porous rock with interconnecting pore spaces to serve as a reservoir; an impervious bed above the reservoir rock to confine oil to the reservoir rock; and proper attitude of the strata. By the latter, Hunt meant the anticline, an upfold or arch of stratified rock in which the beds or layers dip in opposite directions from the crest.

By 1900, geologists had formulated a more inclusive trap theory holding that oil and gas migrate updip in porous strata until stopped under some impervious bed or barrier. The factor that holds the oil or gas in place is called the trap, whether it is caused by a change in the attitude or structure of the rocks or a change in their physical characteristics.

At first, geologists in California, as elsewhere, studied the surface for indications of promising structures to drill. As they found and drilled more prominent structures, farsighted geologists speculated that there might be other structures, as yet untested, whose characteristics did not boldly show on the surface. Specifically, they wondered if the San Joaquin Valley, which had been so productive of oil from fields like Midway-Sunset and Kern River on its borders, might contain worthwhile structures hidden by the alluvial accumulation forming the flat floor of the valley.

To get a better idea of what lay below the surface, oil companies turned to geophysical methods that could delineate drilling prospects. Shell primarily made the early appli-

Rio Bravo discovery well illuminated by torches of surplus gas, being burned to eliminate any possible hazard.
(California Oil World)

cation of geophysical methods to oil-finding in California. The man in charge of Shell's exploration department on the West Coast, first as chief geologist from 1922 to 1928, then as vice president in charge of exploration from 1928 to 1945, was Dr. E. Fred Davis, better known as Fritz Davis. Born in Colorado, he had lived in Houston before enrolling in the University of California at Berkeley, where he received a bachelor's degree from the college of mining in 1910 and a doctor's degree in 1917. Davis operated the seismograph station at U.C. while teaching there before joining Shell as a geologist in the Rocky Mountain area. Shell transferred him to California in 1922 as chief geologist for the Pacific Coast.

The company's first geophysical work in California was with magnetometers. The magnetometer, an instrument that measures the intensity and direction of magnetic forces, had been employed successfully in the search for deposits of iron ore. Shell hoped that differences in the content of magnetic minerals in various California sediments might give clues to structures which underlay valley alluvium.

"We had hoped, for one thing, that we might detect basement highs which would give clues to large uplifts of the type of Signal Hill or Kettleman Hills," Davis recalled. Unfortunately, the instruments the company used seemed more affected by erratic distribution of vivianite in rocks at the surface than by deeper structures the company was seeking. The company gave up early experiments without any successful result.

From out-of-state word came of successful work with the Eötvös torsion balance in Mexico and on the Gulf Coast. The instrument, developed by the Hungarian physicist Baron Roland Eötvös, head of the physics department at the University of Budapest, measured fluctuations in gravity. Over the crest of an anticline, denser rocks lie closer to the surface, and the pull of gravity is greater than on the flanks. Torsion balance observations, made at various points over a given area, provided data from which changes in the pull of gravity could be computed. By correlating gravity data with available geological information, some idea could be gained of the subsurface structure of the area.

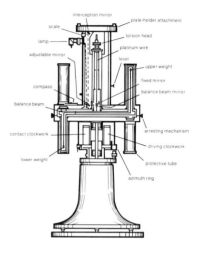

The torsion balance.

"We decided to try one in California," said Davis. "The operation was to be kept a deep secret, and very few people in our own company knew what was going on, nor did they know the people who were engaged in the work."

To help keep the secret, the company picked for the first trial a relatively remote part of the San Joaquin Valley at the Lost Hills field, forty miles northwest of Bakersfield.

The Eötvös torsion balance had a horizontal beam suspended by a delicate wire with two gold weights suspended on the ends of the beam. The two weights were suspended at unequal distances below the beam. The instrument was sensitive to surrounding topography and, in fact, the ground immediately around the balance had to be smoothed off for several feet around the set-up. Several hours were required to complete the necessary observations at a single station, all of which made the operation slow and expensive.

"Nearby rodent burrows or buried stream gravels disturbed the measurements," Davis explained. "Also there were interruptions as the result of a steer coming too close to the set-up during the operation of the balance, and possibly the near approach of a curious geologist in the employ of some other outfit."

The instrument was especially sensitive to temperature changes and, in addition to the insulating jackets which were part of the instrument, it was necessary to protect it from the sun and the cool night air by a small building.

"The hut was quite conspicuous, even when camouflaged," Davis said, "and invariably attracted the attention of passersby and especially the landowner or the caretaker of the property. Our first operator decided to turn aside suspicion by saying that he was engaged in astronomical and geodetic work. One owner of a large cattle ranch approached the hut and found the operator inside and inquired what went on. 'Oh, I am just studying the—' He was interrupted by a, 'I know damned well what you are studying. Don't eat any of this grass,' and the cowman rode off."

In spite of the instrument's apparent simplicity, the calculations were cumbersome. Operators had to be engineers trained to adjust and set up the instruments. While they could measure the records and compute the gradients, their geological interpretations often seemed fantastic.

Unfortunately, results were not good when the company moved away from the known structure of the Lost Hills field. "After a thorough trial in many places in California, it was apparent that while the torsion balance could mark out major structural features, it would give little indication of the secondary folds in which oil occurs," said Davis, who decided that it was a useful reconnaissance method which could not find the oil fields as expected.

Another tool, the seismograph, had shown promise on the Gulf Coast. Originally used to record earthquake tremors, seismic recording instruments had been put to use during World War I by the Germans, who had found adaptations useful in locating enemy artillery emplacements. After the war, Dr. L. Mintrop of Germany adapted seismic methods to oil-hunting. "We learned that in the Gulf Coast various oil companies were using the Mintrop refraction seismograph. We decided that this method might work in California," Davis said.

The refraction seismograph worked by the shock set off by a charge of dynamite on or near the surface. The energy waves emanating from the explosion were picked up by a series of seismometers, or listening ears, placed on the surface of the ground at various points several miles from the explosion point. The seismometers picked up the arrival of the first of the energy waves, making possible an accurate record of their travel time from the point of the explosion. Through soft formations such as sand and shale, energy waves traveled in underground arcs with known velocity between the explosion point and the seismometer. A compact formation would transmit the energy waves at a much faster rate, in effect refracting them much as a prism refracts a light ray. Refracted waves would arrive at the seismometer in an abnormally short time. By setting off and recording explosions at different points, areas could be outlined through which the waves passed faster than normal. Thus salt domes, associated with oil production on the Gulf Coast, could be located.

"With the large charges used, a good-sized cone of earth

A Frank Rieber seismic crew, above, at work in the San Joaquin Valley. In the bottom photograph a shot is fired, left center, and the recording truck, center, makes records for interpretation by geophysicists. (Pete MacMurrough, Petroleum Production Pioneers Collection, Kern County Museum)

221

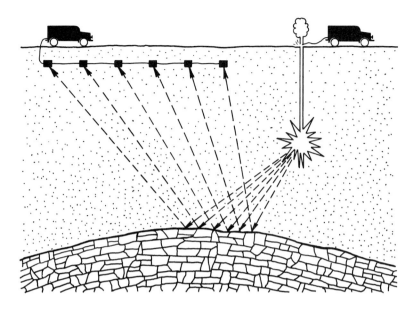

Diagram of seismic explosion and recording truck.

was blown out at each shot point and we had some trouble back-filling these properly," Davis explained. "Winter rains usually caused a settlement of the ground over the old hole requiring a second back-filling job. We finally learned to make a trade with the landowner and paid him to fill up the holes on his own land. Windows were broken and replaced, and some concrete irrigation pipes had to be repaired. With a large crew, which included a claim agent, the operation of a refraction party was quite expensive and, in our hands in California, not especially successful."

Still another technique tried by Shell, but without success, was an electrical exploration system devised by the brothers Conrad and Marcel Schlumberger, of Paris, who would later become famous for their electrical well-logging method. Their system involved setting two electrodes at the end of a long cable, sometimes nearly 50,000 feet long, and passing a current into one electrode so that the current would return through the ground to the other electrode, hopefully giving clues to structure below.

"The operators who brought the electrical instruments to us were, of course, French and naturally they favored French people in hiring their assistants," Davis recalled. "Our various methods of exploration overlapped one another in time so that for awhile we had Weits Cottage Hotel in Wasco pretty well filled up with our crew members, many of them speaking French, some Dutch, some German, and some talking plain Texan or Oklahoman."

Meanwhile another method, the reflection seismograph, had appeared. In this method, a hole several inches in diameter was drilled to a depth ranging from a few feet to several hundred feet, sufficient to penetrate below the layer of loose surface material. An explosive charge was placed at the bottom of the hole, with seismic detectors situated at various distances from it. When the charge was fired, the shock waves reflected off the different rock beds below would be picked up by the seismic detectors and recorded in the instrument truck. The procedure would be repeated at each location in a previously determined pattern. The depth of a particular rock stratum would be determined at each detector station by the length of time required for the shock waves to return to the surface. Seismic profiles could be made in several directions

Diagram shows information contained in a Schlumberger log.

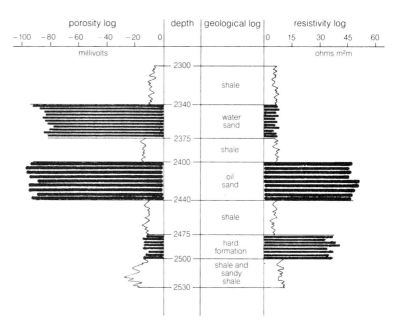

223

across the area, and from this record the depth and dip of beds could be determined and thus the type of structure deduced. The geophones were set out quite close to the instrument truck so that the operation could be carried on more quickly than refraction shooting.

"We learned from our associates in the East that they liked it, and we borrowed a party to try in California," Davis said. "It took us all of six months to learn how to get usable reflections here, but after this we were very favorably impressed. Only light charges of powder were used, which did very little damage to the ground or to irrigation works. We were accused of addling eggs under setting hens, but the charges did not stick."

Shell was one of three companies to shoot an area in the Kern delta near Bakersfield with a reflection seismograph.

Instruments inside recording truck. (Pete MacMurrough, Petroleum Production-Pioneers Collection, Kern County Museum).

Only Shell seemed to perceive a structure. The company leased 6,433 acres from Kern County Land, which years earlier had fenced in a ten-section tract in the area, giving rise to the name Ten Section. On the basis of information gained from the seismic survey, Shell in the winter of 1936 spudded the Stevens A No. 1 on land leased from Kern County Land. Shell launched the wildcat exploratory venture in the face of widespread opinion that the floor of the valley would prove to be nothing more than a wildcatter's graveyard.

On June 2, 1936, the wildcat tested at a rate of 1,200 barrels a day of $60.4°$-gravity condensate and fifteen to eighteen million cubic feet per day of gas from a total depth of 7,880 feet. The discovery not only opened up a new oil field but also gave a name to what would become one of Kern County's best oil sands—the Stevens sand. A Southern Pacific Railroad spur ran through the new-found field. Near the discovery well, the railroad company some years before had laid out a subdivision—still not settled—and called it Stevens Siding. Shell named the wildcat for the siding. When the drill bit found oil sand, the sand became the Stevens sand.

After the discovery, a group of Shell men, including the company president, Sidney Belither, and Fritz Davis, were driving from the Ten Section well to Bakersfield when they saw in a farm field near the road a blindfolded mule plodding a resolute circle around a crude woden contraption that looked as if it had been designed by Rube Goldberg.

The Shell president wanted to know what the mule was doing. One of the men in the car explained that the mule was pumping one of Kern County Land's water wells. The president took a long look at the bucolic scene. "We fixed it today," he said, "so they can retire that old mule."

The Ten Section discovery well had hardly been turned to the tanks before reflection parties worked and reworked the alluviated areas of California. Shell and other companies went to the field with the exciting new tool in search of hidden structures that might contain oil. Though the reflection seismograph would prove a valuable ally in the hunt for oil, it was only a tool—and the final judgment would still be rendered from the floor of a drilling rig.

Eleven miles northwest of the Ten Section field, three miles south of the town of Shafter in what was known as the Rio Bravo area, Superior and Union showed interest and went to work with the reflection seismograph. Bill Keck of Superior had hired geologist Walter English in 1929, and one of the first things English did at Superior, he later recalled, was to start building a seismograph outfit. "I hired several students from Cal Tech as electrical technicians and managed to get a seismograph together which we started to use in 1933." English said that one of their first projects was shooting across the Rio Bravo area. The results looked good, and Superior leased land.

Union Oil, too, liked what it saw at Rio Bravo. Union was not a newcomer to the area. More than ten years before, Union had mapped a topographic structure—a surface high —that looked interesting, and in August, 1924, had spudded in to drill an exploratory well. The hole was dry at total depth of 6,053 feet. Though the venture was unsuccessful, Earl Noble, Union's chief geologist, had not lost faith in the area. The advent of the reflection seismograph gave him another tool with which to appraise Rio Bravo. A seismic survey indicated a structural high—a dome—buried beneath the deep layer of alluvial soil. The seismic high appeared to be about one mile east of the topographic high. On Noble's recommendation, John Church, Union's manager of lands, leased a block of ground from Kern County Land.

At Rio Bravo, Union told its division drilling foreman to take out a rig and drill to 8,500 feet for a look at the Stevens sand, the same horizon that had yielded discoveries at Ten Section and Greeley.

The foreman was Jack Reed, who more than thirty years before had begun his career with the company driving a "hurry-up" wagon around the Purisima lease at Lompoc. At Rio Bravo he had a rig quite different from the cable tool outfits he had used in the early days. This rotary drilling rig had a towering steel derrick and big steam boilers to furnish power. Moreover, there were men filling jobs that had not existed in cable tool days. The engineers, Carl Steiner and Dudley Tower, would be in charge of the drilling program such as proper pipe sizes and the depth at which pipe should

Earl Noble, left, chief geologist, and Jack Reed, division drilling foreman, studying a core from Rio Bravo discovery well. (California Oil World)

be cemented. Reed and Frank Guess, district drilling foreman, would oversee the actual drilling, coordinating their efforts with the work of the engineers. The geologists—Earl Noble, Louis Waterfall, and E.R. Atwill—would evaluate the exploratory well as drilling progressed. The paleontologists, Glenn Ferguson and Aden Hughes, would look for "bugs" to tell the geologists what formation the well had reached. This sophisticated array of talented and well-trained experts contrasted sharply with the determined, on-the-job training of Carl Baker and the other drillers who had pioneered the valley oil fields a generation earlier.

On March 29, 1937, Union crews spudded in to drill the Kernco No. 1–34. Four days later, they stopped to cement a string of $13\frac{3}{8}$ inch casing at 1,143 feet. The casing protected ground water from damage and served as an anchor for the blowout prevention equipment. The blowout preventers offered much more of a safeguard than any that had existed in cable tool days.

Twelve days after spudding in, Union began to core, using another technique that had not existed in cable tool days. Just below 4,500 feet, they picked up firm, dense, massive, gray shale. In cores, paleontologists identified common fish scales

227

and spines, which helped them recognize the formation as the Etchegoin, a producing formation in other San Joaquin Valley fields but dry at the Rio Bravo wildcat.

Crews cored almost continuously to a depth of 6,600 feet, reaching that depth one month after spudding in. At that point, the company ran a Schlumberger Electric Core Survey to gain information that would help correlate this wildcat well with others like the Shell and Standard discovery wells, and also to help to tell them what they had passed through. The electric log was another technique that had not existed thirty years before.

Sixty-nine days after spudding in, crews had reached a depth of 8,682 feet. They cored a four-foot interval, recovering firm, generally massive, green-brown to green-gray shale with one slight show of oil in a porous streak. In the core they also found an almost perfect leaf imprint, which paleontologists noted with interest. They were in the Stevens zone.

Jack Reed recalled that "we didn't have anything but high pressure salt water and a little gas."

It was painfully obvious that Union did not have another Ten Section or a Greeley. The company's decision-makers debated, finally deciding to go down further. In early June, with the hole down to 8,805 feet, Union ran another Schlumberger Electric Core Survey, but still found no pay sand. The rotary table turned to the right, and the bit ground out more hole.

"We took it on down to 10,200 feet," Reed recalled, "and then the top men in the field department came around and said they were going to shut it down."

The well was rapidly approaching a depth beyond which no one had ever found oil. Five months before, Associated Oil had completed the McGonigle No. 12 in the Ventura field as the world's deepest producing well, getting an initial production of 536 barrels a day from 10,569 feet. However, this was not impressive production considering the depth of the well and the 338 days it had taken to drill it. Some engineers argued that such deep wells would not pay drilling costs, even if oil were found.

"I argued and argued against shutting the well down," Reed said.

Those opposed to going on argued that any sand found at that depth would be so tight from the weight of earth resting on it that it would not produce anything.

"They said they knew that so much earth was setting on this that if they went down and hit the Vedder it wouldn't be productive," Reed recalled. The Vedder, a sand that occurred below the Stevens, had been found productive in fields on the east side of the San Joaquin Valley, where sands were much shallower than on the floor of the valley. "They argued there'd be no permeability or porosity at all because there was so much overburden," Reed explained.

In oil terminology, porosity is a measurement of pore space in reservoir rocks, while permeability is an expression of the ease, or lack of ease, with which fluid may pass through the rock. Both are key factors in recoverability of oil from reservoir rocks.

"I told them we hadn't proved it yet. But anyway, we argued and argued and finally we went on down," said Reed.

On July 17, with the hole down to 10,310 feet, the crew was pulling drill pipe from the well when the string of pipe stuck at 7,938 feet. The crew pumped in sixty barrels of water and attempted to reverse circulation to try to free the pipe, but the formation absorbed the water, leading to a circulation drop, and the effort was unsuccessful. Instead of an oil discovery, Union had a fishing job at Rio Bravo.

The fishing job coincided with disturbing news elsewhere. At Greeley, four miles to the southeast, Standard Oil was getting a great deal of water in the first follow-up to its discovery well there. The company had two rigs at work, but the commercial significance of the Greeley field remained a puzzle.

At Rio Bravo, Union conditioned the hole to prevent it from sloughing, or having its sides fall in. This was accomplished by adding weighting material to the drilling mud, a technique developed after World War I.

The stuck drill pipe was $3\frac{7}{8}$ inches in diameter. Union ran a string of $1\frac{1}{2}$ inch drill pipe with an inside cutter on it through the stuck pipe. They cut the pipe at 7,820 feet, and the upper portion was pulled from the hole. This endeavor left a 115-foot "fish" in the well.

Ponderous equipment helped drill the discovery well at Rio Bravo. (California Oil World)

The first attempt to recover the fish by "washing over," or encircling it with a pipe shoe of larger diameter, failed when the washover shoe went alongside the fish instead. The crew ran a string of pipe into the hole with a releasing socket, which they managed to work over the fish several times. They took a 132-ton strain but the slips in the socket would not hold the fish. They went back in with wash pipe and washed by the fish to a depth of 7,934 feet. They ran back in with a releasing socket and back-up equipment, took hold of the fish, and this time jarred it loose, recovering the fish ten days after the fishing job began. The tools and techniques they used to successfully resolve the fishing job were far advanced from the equipment available in cable tool days.

Another electric log survey was finished. On August 5, some four months after spudding in, Union cemented a string of $6\frac{5}{8}$ inch casing at 10,248 feet. The company ran a formation test from 10,250 to 10,254 feet, getting a strong blow of gas at the start, which decreased to a faint and intermittent blow. The well died after twenty minutes. On a test from 10,248 to 10,318 feet, the well failed to flow. When the crew pulled pipe, they recovered nine stands of gassy mud. Union drilled and cored ahead to 10,383 feet, tested again and recovered gassy water and mud. They cored ahead to 10,408 feet, tested again and got oily, gassy mud.

In mid-September, the crew cut a core that took the hole from 10,692 to 10,700 feet. Two feet from the bottom, there was abundant pyrite, better known as fools' gold.

Time seemed to be running out, but crews continued to core ahead as the company went on with the deep probe in spite of the lack of encouragement.

On October 13, crews cored the interval from 11,236 feet to 11,255 feet. When they came out of the hole with the core, they found oil sand with a good gasoline odor. They were elated.

"You could take a piece of it about a foot long and wrap tinfoil around it and stick it in water and blow through it endo," Reed said. So much for the theory that sands at such depths would be too tight to produce.

For fifty-five feet, crews cored gradually darkening sand. Solvent tests on core samples gave excellent showings. Two days after picking up oil sand, the company bottomed the well at 11,302 feet.

Since a possibility existed that the well might be taken deeper, the setting of liner at a depth in excess of two miles presented some difficulty. Liner, normally set opposite the oil zone to keep sand in place while allowing oil to enter through perforations, would have to be removed if the hole were deepened. Engineers settled on an aluminum liner as the answer. If they decided to go deeper, crews could easily drill out the aluminum liner. Normally, engineers would have set a pre-perforated liner right down into the producing sand, but in this case they decided to use a solid liner, setting the bottom twenty feet above the sand formation, to protect that portion of the hole below the $6\frac{5}{8}$ inch casing at 10,248 feet and the top of the oil sand. In the vocabulary of the oil man, they completed the well "barefoot."

"When we brought in the well," Reed recalled, "it came in and flowed at the rate of 30,000 barrels per day for four and one-half hours. We filled up everything we had around there." Glenn Ferguson, the paleontologist, recalled that "when they first turned it to the tanks, the oil boiled up like a spring and just kept coming. I wish I had a picture of it," he said.

The well was the discovery well for the Rio Bravo field. On November 4, 1937, it flowed at a restricted rate of 2,400 barrels a day of high gravity oil through a $\frac{40}{64}$ inch "bean" or "choke," a restrictive device designed to slow the rate of flow.

The well proved that profitably productive oil sands existed at previously unexplored depths, that the technique of drilling had progressed sufficiently to probe those depths, and that in consequence much future drilling might likely be devoted to drilling deep wells.

In the wake of the discovery, the development of the Rio Bravo field marked the first extensive use of 172-foot and 178-foot steel derricks together with fourteen-inch by fourteen-inch steam engines for driving modern, fully enclosed, double chain, oil bath drawworks with independent, steam-driven rotary tables. Boilers of 350 pounds working pressure and steam pressure became the common sources of power. Rock bits became the accepted type of bits, and operators turned to high rotary table speeds up to 350 to 400 revolutions-per-minute and the use of a large number of drill collars to guarantee the drilling of straight holes. In contrast to the 221 days required to complete the discovery well, fol-

Towering steel derricks exemplify changes in drilling techniques from pioneer days; Kettleman Hills, 1930's. (Standard Oil Company of California)

low-up wells were completed in as little as sixty-one days, with several making over 10,000 feet of hole (including cementing of surface casing) in the remarkable time of thirty days.

At 11,302 feet, Union's discovery well was the first well in California to produce oil from below 11,000 feet. It was also —at the time—the deepest producing oil well in the world.

"I was pretty proud of it," Jack Reed said.

Index